EVOLUTION
IT NEVER HAPPENED

Dr. Patrick I. Amadasun Ph.D., DM.

First Edition

US Paper Warehouse, Inc.
10990 Galen Place
Johns Creek, Georgia 30097
USA

Evolution:
It never Happened

eBook ISBN: 978-1-7355241-6-0
Paperback ISBN: 978-1-7355241-7-7
Hard Cover ISBN: 978-1-7355241-8-4

Library of Congress – Registration No.: TX 4-919-623

eBook ISBN: 978-1-7355241-6-0
Paperback ISBN: 978-1-7355241-7-7
Hard Cover ISBN: 978-1-7355241-8-4

Printed in the United States of America

Dedicated to all peoples of the world, especially Christians that endeavor to seek scientific truths which were garnered from evidence based research. Creations in the universe are perfect and endearing.

About the Author

Dr. Patrick I. Amadasun Ph.D., DM, has a background in Biology and Medicine. He did his Doctoral studies in Finance, Economics, and Management at Georgia State University and University of Maryland. He is a life member of Phi Kappa Phi and a member of the Academy of Management. He was former Chairman of First Continental Insurance and seat on the Board of several Corporations. He has been a consultant for numerous corporate entities, including a nation. He is an erudite scholar and has authored several works of scholarship.

Table of Contents

CHAPTER I:

INTRODUCTION – BEFORE AND BEYOND DARWIN

"False facts are highly injurious to the progress of science, for they often endure long; but false views, if supported by some evidence, do little harm, for every one takes a salutary pleasure in proving their falseness". Charles Darwin

There have been various hypotheses and theories proposed by different scientists and religious organizations with the intent to explain the origin of life; also, how living organisms evolve and interrelate on our planet. In this book, brief information on the various thoughts on evolution was presented in this chapter and forms the basis for the direction taken in this text that enables the illumination of a novel evolution theory based on some interesting research findings.

It follows, therefore, that the purpose of this book is to ascertain if the various existing evolution theories and hypotheses have any factual and scientific basis for their conclusions; and, present an alternative narrative that was derived from evidence base research. To this end, small experiments were carried out and presented in the following chapters. The conclusions from these experiments will forever change biology and evolutionary thoughts; and, will completely disprove Darwin and other evolutionary theories. It will cause a brouhaha in the scientific world – that, evolution never happened. Furthermore, the research findings formed the basis for a new evolutionary theory created by the author and referred to as Amadasun's Theory of Absolutism (see Figure 1.1).

Figure 1.1: Picture of the Galaxy.

Firstly, let's delve into the beginning of life on our planet and examine the various evolutionary thoughts in the scientific world. The origin of life has been so much shrouded in mystery that most scientists, including biologist, astronomers, geologists, etc., and different religious groups, have developed different theories in an attempt to explain this phenomenon. Religious organizations have hypotheses that were generated from the Bible which support their underpinnings in their attempt to explain the existence of life in the universe as written in the book of Genesis. Their theory was that life originated in this universe within seven days by a creator.

"In the beginning, God created the heavens and the earth. Now the earth proved to be formless and waste and there was darkness upon the surface of the watery deep; and, God's active force was moving to and fro over the surface of the waters." Genesis: Chapter 1.

This belief led to the creationists' theory that the origin of life was by divine ordination. But, then, is there a basis for this belief? Are there any scientific evidences to back this belief? (see Figure 1.2).

Figure 1.2: Picture of Blue and Black Milkyway.

Scientists, generally, have disagreed with the creationist science because the belief is not supported by scientific evidence that could be tested by: (a) materialism – scientific explanations that are grounded in material cause and do not violate natural, law; (b) testability – that the predictions made cannot be tested against the material world; (c) that their belief is unchallengeable, even in the face of new scientific evidences that prove otherwise.

In other words, the creationist theories are hearsay that is not supported by scientific facts. But, could scientists be wrong? Are there new scientific facts that support the Creationists' beliefs? Let's keep reading as some interesting research findings are presented in this book.

The belief of most scientists is that the origin of life can only be explained and comprehended in the context of evolution. This theory expressly states that all organisms are related through a descent from a common ancestor. This theory has become the core foundation of biology. It relates the animal kingdom in terms of physiology reproduction, biochemistry, and other functional systems of the organism as being inter-related and demonstrates an evolutionary trend on how life came into existence.

Is this theory correct? Has this been fully proven to be scientifically factual?

Regardless of the philosophical beliefs of the various theories, the Darwin's evolutionary theory has become the dominant scientific theory taught in all educational systems worldwide as being the theory that best explain life in this universe.

The geneticist, Theodosinus Dobzhansky, aptly supports this view in his declaration that, "Nothing in biology makes sense except in the light of evolution".

If indeed we all evolved from a common ancestry, who or what then, was the first ancestor of life? How did that ancestor come to life? Was it created, or, did it create itself? Was it accidental, or was it divinely ordered?

The "Big Bang Theory" tried to explain this phenomenon in terms of organic matter by focusing on the planet Earth through the examination of its geological data which reveals that the earth is about 4.6 billion years old. The theory suggest that the earth was formed from the coalescence of gases and debris from exploding supernovas, and started as a molten core of nickel and iron surround by silicate rock. Scientific and geological data stated that the earth was a barren planet surrounded, or filled, by toxic, reducing gases, ammonia, methane, and hydrogen.

Interestingly, there seems to be a correlation of this early beginning from the "Big Bang Theory" and the explanation presented in the bible. In Genesis, the first book of the bible, it is quoted as such: "_ now the earth proved to be formless and waste and there was darkness upon the surface of the watery deep". It appears that the "formless and waste" referred to in this passage of the bible supports the creationist theory and show consensus to the geological data of the "Big Bang Theory" which explains that the earth, initially, was barren, filled with toxic, reducing gases. What a significant correlation! Another correlation was again shown in Genesis. The passage continues: "and God's active forces were moving to and fro over the surface of the waters___." The "Active Forces" here show correlation to the array of steady bombardment of cosmic and ultraviolet radiation of the initial planet, earth, as expressed in the "Big Bang Theory".

This theory further explain that after the earth's cooling, there were a lot of chemical and physical reactions taking place and the molecules formed did combine to form organic compounds and later form the living cell! Is this theory conclusively factual?

Could life have evolved, or, originated this way? Why was this combination so orderly? Why was this combination so consistent? Why are there no different forms of life that are fully formed from other chemical elements instead of consistently organic elements? Why have we not fully observed such phenomenon again in the universe, thereby, creating another planet with live organisms on it? It has been over 4.6 billion years since this occurrence took place; why don't we see, or observed more of this phenomenon on a yearly, monthly or daily basis? If this phenomenon is right, why don't we have zillions of planets filled with lives? (See Figure 1.3).

Figure 1.3: Picture of Extinct Dinosaurs.

If we apply the law of thermodynamics which states that all closed systems–whether chemical, physical, biological, and other such systems–tend over time, to become increasingly disordered, or become entropic, then: Why does the living cell, a biological system, seem to violate this law? Why are the living cells so orderly?

Could these theories on the origin of life be faulted by the preponderance of available scientific evidences that suggest that the orderliness found in living organisms and in the planet earth does not show consonance with the "Big Bang Theory"? The dissonance observed in these theories cannot be mitigated by deductive or, inductive reasoning that are backed by observed facts or, could be inferred from certain observations that should have happened, but never occurred?

It therefore, follows that the origin and diversity of life in this universe has not been fully explained by science in a manner that will finally remove the controversy or the conundrum surrounding this phenomenon.

If we accept that the origin of life is as stated; how did it evolve? And, if indeed there was evolution, how does it function? How is it organized? Why does it take place? What role does it play in human originations? Let's briefly probe into some various evolutionary theories and see if they were able to provide answers to this conundrum.

In 350 BC., Aristotle broached the Scala Naturae theory and suggested that species are incapable of changing and there is no mechanism for evolutionary change. Aristotle may have been proven wrong on evolutionary change; however, could he have been right in his suggestion that species are immutable? In 1305, Ramon Llull adopted the Great Chain of Being theory which supported Aristotle through Christian lens.

Caspar Wolf proposed the Vitalism theory in 1759 which has a foundation in various theories from Ancient Egypt. He believed that species can change and the mechanism of change is the life force in the embryo. This theory was quickly discarded by biologists as untenable in 1828. Similarly, Asa Gray advocated the Theistic Evolutionary theory in 1871. He believed that species can change and such changes were brought about by various deities. This theory was also rejected by biologists in 1900 for lack of scientific proofs.

The theory of Orthogenesis was first advanced by Karl von Baer in 1859 and later supported by Pierre Teirhard de Chardin in 1959. The theory suggests that species can change and the mechanism for change is through "purposeful creation" and the "inherent progressive tendency" which has a foundation in spiritual theory. Biologists also discarded this theory. But, is it possible that these scientists could be on the right path?

Lamarckism by Jean-Baptiste Lamarck in 1809 was a theory that agrees that species do change and mechanism of change was by the use and disuse of inheritance of acquired characteristics. This theory has been rejected by biologists. Likewise Catastrophism by Georges Cuvier in 1812 was rejected due to the fact that it lacks scientific fact and rejects changes in species.

D'Arcy Thompson in 1917 proposed the theory of Structuralism which supports species change that is brought about by self-organization and physical forces. It marked a beginning of the view that cellular self-organization could explain evolutionary change. Interestingly, before Thompson, Geoffroy Saint-Hilaire in 1831 suggested the theory of Mutationism which supports species change that occurs through large mutations in the sudden production of new species under environmental pressure. The theories suggested by these scientists seem to have scientific value and have not been fully discarded (see Figure 1.4).

Figure 1.4: Picture of Progressive Evolution.

In 1859, Charles Darwin advocated Darwinian Evolutionary theory which supports species change that is brought about by natural selection, but lacks mechanism of mutation brought about by genetics. In 1968, Motoo Kimura proposed the National

Theory of Molecular Evolution which suggested that species change was brought about by genetic drift at the molecular level and supports Darwin's natural selection at the higher or phenotypical level. Consequently, the domineering evolutionary theory that is supported by biologists is the Darwinian Evolutionary theory.

But, then, is Darwin right? Let's explore Darwin theory more and see if there are gaps and fissures in this theory.

Charles Darwin played an important role in trying to explain the theory of evolution. At the time he wrote his book, evolution was a very controversial subject. It still is. During that period, Darwin's findings appear to be very logical and realistic. These findings became the foundation of current evolutionary theory which helped in shaping the evolutional trend of life diversity as adopted by most biologists, and as taught in all schools and colleges.

Darwin's book, "The Origin of Species by Natural Selection, or the Preservation of Favored Races in the Struggles for Life," was written after his trip to the Galapagos Islands. He tried to explain evolution as being a characteristic of the living world. He fully stated that all organisms are descended from a common ancestor, and we all belong to the same genealogical network. Secondly, he stated that the process of evolution, or, descent with modification, is constantly taking place from generation to generation. According to Darwin, the motor of change in evolution was the principle of natural selection. In other words: (1) since resources cannot accommodate the potential growth of population, there will be competition for resources, or, "The struggle for existence"; (2) the characteristics of the individuals are heritably passed on to their off-springs with the characteristics most important for survival being given more prominence; and, as a result of these, the overall composition of the population is being changed from generation to generation with new features being developed, and some old ones becoming extinct. This situation result in gradual evolution taking place.

But, then, are Darwin and other evolutionist theories accurate? Did all organisms descended from a common ancestry and did they all belong to the same genealogical network? Are most mutations occurring favorable to the organism? Does it not appear that most mutations have in fact been injurious to the organism? Are most mutations occurring peripheral, or, do they also affect internal tissues and organs?

Generally, mutations are lethal to the organism, and are not generally helpful to organism. Because of these lethal mutations, it is more difficult for organisms to survive mutations in large numbers that will result in favorable evolution with regards to living organism.

Did evolution really occur? How did it occur? Are there enough surviving organisms, with favorable adaptations from generation to generation, to evoke an evolutionary trend?

In continuation of the support of Darwin's Theories, most scientists do believe that, without a doubt, evolution did occur, but, natural selection may not be the only force responsible for this organic change, and, that other material forces may come into play.

The theory of evolution has been one of the most controversial theories ever proffered due to its affective effect on humanity, religion, and association to other forms of life that people and the church found offensive, including reference to our common ancestry with apes. This philosophy was controversial in Darwin's time and it still controversial today.

Can Darwin be correct? Are the evolutionary theories challengeable? Mendel's work on the Law of Inheritance shed more light on the functions of the cell and how genes are the main organic transfer of cellular information from an organism to its offspring.

Initially, Mendel's work was not fully recognized, and it appeared as if it challenged Darwin's theory of evolution. Eventually, it was observed that Mendel's breeding experiment with garden peas, and using "factors," was referring to genes, which will further shed light on how variation (the mechanism of evolutionary changes) occurs. Additionally, the works of James Watson and Francis Crick helped resolve the structure of the DNA, and most importantly, during mitosis, or meiosis, the diploid chromosomes state or, haploid chromosomes state in the gamete cells, could only result in the parent DNA, making copies of itself (see Figure 1.5).

Figure 1.5: Picture of Genes.

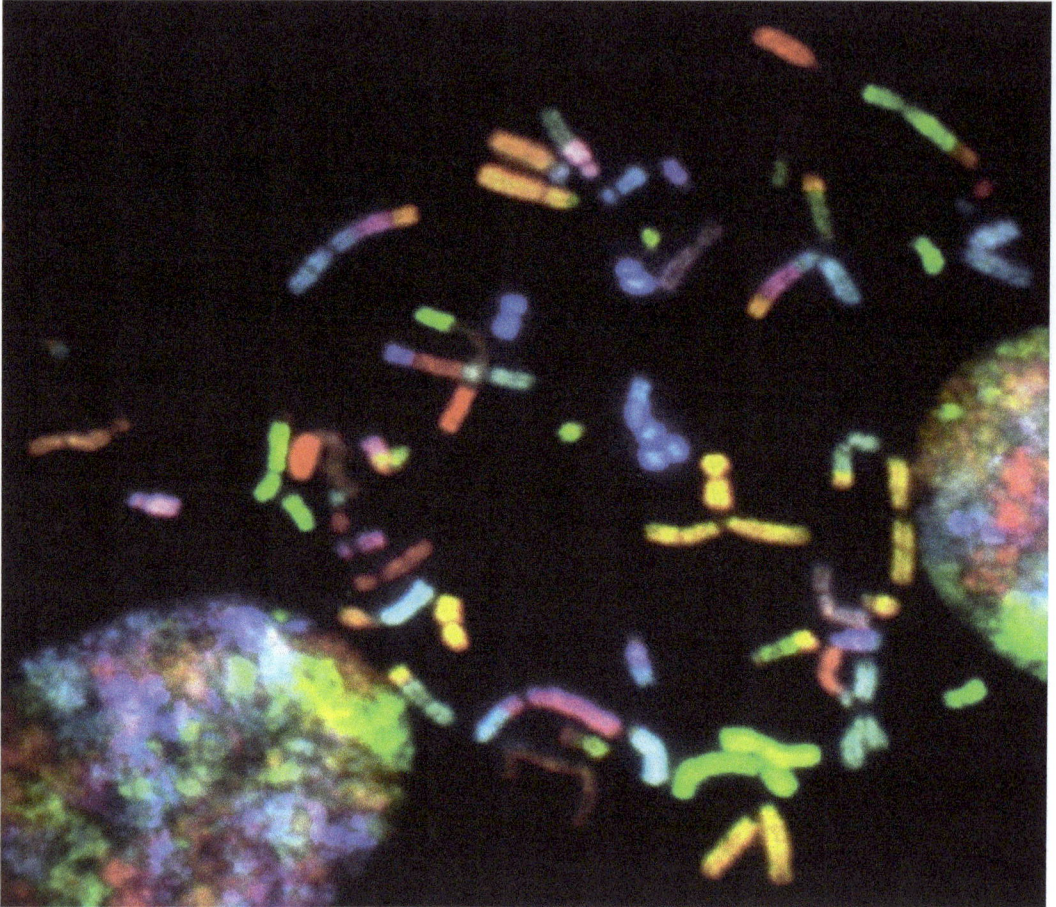

But, it should be noted that the copies were not directly identical copies, but complimentary copies.

Further works revealed that the DNA composition and structure of nucleotide do form molecular words, or, code that the cell interpret in carrying out cellular function. These triplet codes were responsible in coding messages for producing specific protein and enzymes responsible for cellular functions and metabolism. A change in a single letter of these triplet codes could result in the production of different amino acid, and change the structure of the protein, hence, creating a point mutation. A change in the structure of the DNA by addition, or, deletion of a nucleotide, is referred to as frame shift mutations, and these could be more deleterious to the organism.

Observations and experiences have shown that chemicals, sunrays, ultra-violet rays, x-rays, and nuclear, or, radioactive elements, which are referred to as mutagens, could cause spontaneous mutations in organisms. It is these types of mutations that have been responsible for adaptations and variations that ultimately lead to evolution.

But, then, are all mutations favorable to the organism? Which cellular activity directs the DNA to produce proteins by using the triplet code when needed, and where? How and what should the DNA structure be composed of? How does the DNA recognize these codes and directly respond to them? Are these mutations sufficient over time to create an evolutionary trend?

For example, a mutation what involves the chromosomes number being less than 23 in humans have created serious medical, life threatening conditions. Down syndrome is a result of a trisomic condition with the human chromosome being 21. Similarly, the Turners Syndrome, the Klinefelter' syndrome, and the Philadephia syndrome, just to name a few, are all mutable conditions where the human chromosome number of 23 has been reduced, and are all life threatening situations. There are many medical problems created by mutations, and not all mutations have been favorable to the living organisms.

Could mutations in the genes have been responsible for evolution? If mutations eventually succeeded in creating evolution, were there any surviving organisms to carry on the process of progenity? Could variation have resulted in more termination of the organism's offspring? Could the struggle for the survival of the fittest has resulted in extinctions, instead, of breeding a favorable race? Is Darwin's theory of natural selection peripheral, and does not affect the gene, or, internal cellular structure?

The evolutionists do believe that evolution takes place gradually over time, and it operates by modifying what is already available, and do not build organism from scratch.

Then, who built the first organism? How did the first organism came into existence or, from where or what did it evolve from? Should scientific facts or theories be ambiguous? Shouldn't scientific facts be finitely certain?

The living cell could be compared to a computer. The genes on the chromosome within the cell could be compared to the microchip in the central processing unit. The genes on the chromosome within the cell's nucleus do control cell functions, metabolism, and survival; and, is the brain behind all the cell's activities. Without the gene, the cell is dead, lifeless, useless, chaotic, and non-functional. Similarly, a computer without the microchip, or, controlling software, is just dead and non-functional hardware. Man created computer and the software. It is therefore possible, that the cell and the genes could not have been functional without a creator? Could the Creationist theory on evolution be right? (See Figure 1.6).

Figure 1.6: Picture of an extinct Dinosaur structure in a Museum.

Is there, therefore, a God? Could the bible be real? Did the universe become formulated through a big bang? Are Darwin's theories of evolution correct? Is man an accidental creation? Did man evolve from ape? Did the animal kingdom evolve from a single cell? How did plants come into existence? Why are you white? Why are you black?

Why do you exist? Will your children be heterosexuals, or same sex individuals? Will your children grow up to be intelligent, or stupid? Could you live to be over 1,000 years old? Can science now permanently stop and cure all sicknesses and diseases, including cancer, AIDS, heart diseases, etc.? Is true human cloning possible, with regards to Dolly the lamb cloning? Why is there so much order in cell formation? Why are the DNA (Gene), chromosomes, nuclei, and cellular structure so orderly formed and organized? Why is the information on the DNA so genetically coded accurately to produce certain functions of the body correctly all the time? Why? Why? And, Why? What directs the DNA to produce different proteins through the triplet code? How does the DNA know, or recognizes, when a protein is needed in a part of the body, and where? Despite mutations, the gene has never failed to give every individual organism its own unique characteristics, which completely differentiates it from other organism–like its own unique voices, its own unique set of fingerprints, its owns unique facial recognition, its own unique mannerisms, its own unique colors, etc. Why is every individual a unique organism recognizable from others? Is it possible that beside the DNA, that a new cell organelle, not yet discovered, is exerting some control? Could this be the mastercode as discovered from findings revealed in this book?

All these questions and more will be answered in this book by scientific and genetics facts gathered through experiment and observations that have spanned over 10 years. Perhaps, the most important factor about science is its dynamism–ever changing and continuously challenging with new information from new experiments and observations and these are constantly enriching our lives.

Consequently, the focus of this book is to challenge the veracity of evolution theories, especially Darwin's. In order to achieve this, a lot of interesting observations were recorded from experiments carried out on different species belonging to different phylum of the animal kingdom. A lot of these findings will have effect on how we view our existence in the universe, and will eventually enrich our lives in terms of the health of mankind. Accordingly, the philosophy of this book is grounded on the fact that scientific evidences should be seen in its entirety on how it affects life in the universe. The study, therefore, focus on genetic functioning and behavior, and how its mutation and recombination could affect evolution and the original of life in this universe.

Therefore, it should be noted that for there to be a true evolution, there has to be a move from a species to a genus, to a family, to an order, to a class, to a phylum, to a subkingdom, and to a kingdom, or, from one of these levels to another. In the conduction of this study, observations were made to examine the genome, or gene size, the gene number, the chromosome number, and the cellular organelles of different classes of animals, and to examine the evolutionary trend and if any that were observable. The effect of mutagens where observed on different classes of the animal kingdom to see how it affects the evolution of the organism.

Subsequently, the geological time scale of evolution was constantly referred to, and will be used in our exploration of the origin of life, and its evolution. This book will provide the scientific and evidential proof, or, scientific and genetic facts that will disprove the theories of evolution, including Darwin's and, as such, develop new and factual theories that will tend to explain life origination, existence and sustenance in our universe. Hence, the theory of Absolutism or perfect creation is proffered.

Suggested Reading

Ahern, Kevin, Biochemistry and Molecular Biology, The Great Courses, The Teaching Company, 2019.

Audesirk, Teresa and Audesirk, Gerald, Biology, Life on Earth, 5th Ed., Prentice-Hall, 1999.

Borman,Stu "Protein Factory Reveals Its Secrets", Chem & Eng News: 85(8) 2/19/2007, p13-16.

Burton, Alan C., "Physiology and Biophysics of the Circulation", Chicago, Yearbook Medical Publishers, 1965.

Chernecky, Cynthia, et al., ECG's and the Heart, W. B. Saunders, 2002.

Enger, Eldon D. and Ross, Frederick C., Concepts in Biology, 10th Ed., McGraw-Hill, 2003.

Grauer, Ken, A Practical Guide to ECG Interpretation, Mosby Year Book, 1992.

Guyton, Arthur C., Basic Human Physiology, W. B. Saunders, 1971.

Hickman, Cleveland P., Roberts, Larry S., and Larson, Allan, Integrated Principles of Zoology, 9th. Ed., Wm C. Brown, 1995.

Karp, Gerald, Cell and Molecular Biology, 5th Ed., Wiley, 2008.

Kim, Y., Coppey, M., Grossman, R., Ajuria, L., Jimenez, G., Paroush, S., Shvartsman, S., Current Biology: 20, 3/9/2010, p1-6.

Levy, Charles K., Elements of Biology, Addison-Wesley, 1982.

Lodish; Harvey with Berk, Matsudaira, Kaiser, Krieger, Scott, Zipursky and Darnell , Molecular Cell Biology, 5th edn, W.H. Freeman and Company,2004.

Matthews, C. K., van Holde, K.E., and Ahern, K. G., Biochemistry, 3rd Ed., Addison Wesley Longman, 2000

Moore, R., Clark, W. D., Kingsley, R. S., and Vodopich, D., Botany, Wm. C. Brown, 1995.

Nave, C. R. and Nave, B. C., Physics for the Health Sciences, 3rd Ed., W. B. Saunders, 1985.

Nelson, Philip, Biological Physics, W. H. Freeman, 2004.

Shier, David, Butler, Jackie and Lewis, Ricki, Hole's Human Anatomy and Physiology, 11th Edition, McGraw-Hill, 2007.

Thibodeau, Gary & Patton, Kevin, Anatomy and Physiology, 3rd Ed., Mosby, 1996.

Tuszynski, J. A. and Dixon, J. M., Biomedical Applications of Introductory Physics, Wiley, 2002.

Yockey, Hubert, "Information Theory, Evolution, and the Origin of Life", 2005.

CHAPTER II:

PROTOZOA—"IS THERE LIFE HERE—?"

"Scientists who go about teaching that evolution is a fact of life are great con-men, and the story they are telling may be the greatest hoax ever. In explaining evolution, we do not have one iota of fact." Dr. T. N. Tahmisian

Protozoans are unicellular organisms with over 65,000 species. Generally, they represent the simplest form of living organism within the animal kingdom. It is believed that their first existence was around 3,600 million years ago during the Precambrian period. Evolutionists refer to these one-called organisms as the forerunner of all living organisms (see Figure 2.1).

Figure 2.1: Picture of Unicellular Organisms.

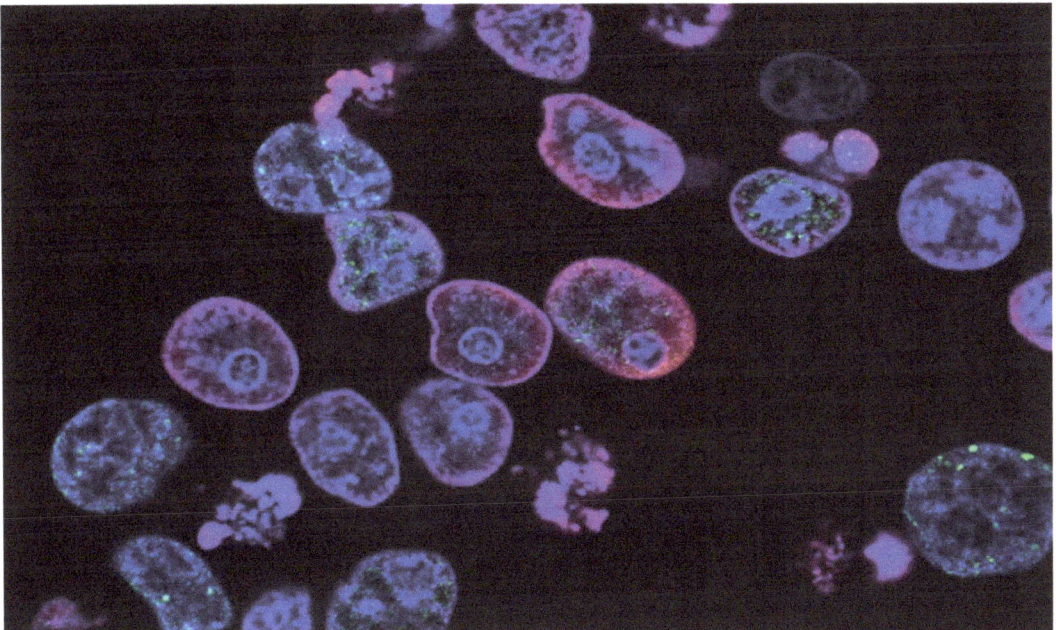

In our study of protozoans, we chose the ciliate paramecium. The experiments carried out were mainly exposure to mutagens like x-rays, ultra-violet rays, and radioactive chemicals, over long periods of time. The control culture was not exposed to any of these mutagens. Our findings will be discussed under different sub-headings as it relates to the metabolic functioning of paramecium. We will also discuss our observations with regards to differences in behavior, genes, mutations, and possible changes in observable evolutionary trend.

BODY COVERING AND MOVEMENT:

Paramecium does have ciliary organelles for locomotion. These unicellular organisms are generally asymmetrical, and do have their shapes maintained by complex pellicles, which are living external coverings of solid cytoplasm. Most of the external surface is covered by longitudinal cilia, or bristles. The cilia are offshoot from basal granules, called kinetosome. These protozoans do move about by the synchronized beat of the cilia. This pattern of movement causes the organism to swim in a spiral manner.

When the organism was exposed to low density ultra-violet rays and x-rays, it quickly swam away from these rays. The organism was then exposed to rays from all the directions. Initially there was avoidance from one end to another as the intensity of the rays were gradually increased, there organism gradually ceased movement and became dormant at the bottom of the petri-dish. Attempts to prod the organism to move became futile. The organism was left in this condition with the rays directed at it fully for a month. It was finally taken out and moved from these rays. It appears that the organism remained permanently comatose.

NUTRITION:

Paramecium does possess a mouth, or the cytosome which opens into a short canal, called the cytopharynx. This canal leads into the cytoplasm where the food vacuoles are formed. Paramecium generally feed on micro-organisms within the surroundings water like bacteria and other ciliates. The beating of the cilia creates currents that drive food particles to the cytosome and down to the cytopharynx. These food particles do collect into a food vacuole at the end of the cytoplasm, which then breaks away after reaching a certain size. These vacuole break away from the cytopharynx and circulates within the cytoplasm. Lysosomes, which carry the digestive enzymes, fuse with the food vacuoles, and active digestion takes place within the food vacuoles. The digested food products are expelled through the membranes into the surrounding cytoplasm. The waste food products, which remain in the good vacuoles, move towards the posterior end to the cell anus, or cytoproct, where the wastes are expelled, and the vacuoles disappear.

Under microscopic observations, paramecium was seen to feed actively. When exposed to mutagens, especially ultra-violet and x-rays progressively, with the duration and intensity gradually being increased, it was observed that the organism gradually ceased all feeding activities, and remained semi-comatose. When the effects of the rays were completely diminished, the organism commenced normal active feeding. However, exposure to a higher intensity of the rays, for a longer duration, led to the death of the organism.

REPRODUCTIONS:

Reproduction in paramecium is usually by conjugation. The two organisms come together and adhere, and there is fusion of their cytoplasm. The macronucleus is generally absorbed, and the micronuclei undergo two meiotic divisions. In each of the organism, only one of the haploid micronuclei survives, with the other three being re-absorbed. The surviving haploid micronucleus in each organism will then undergo a mitotic division to form stationary and wandering micronuclei. The wandering or mobile micronuclei from each organism moves towards the other respective organism and fuse with the corresponding stationary micronuclei. (see Figuere 2.2).

Figure 2.2: Picture of Unicellular Organisms.

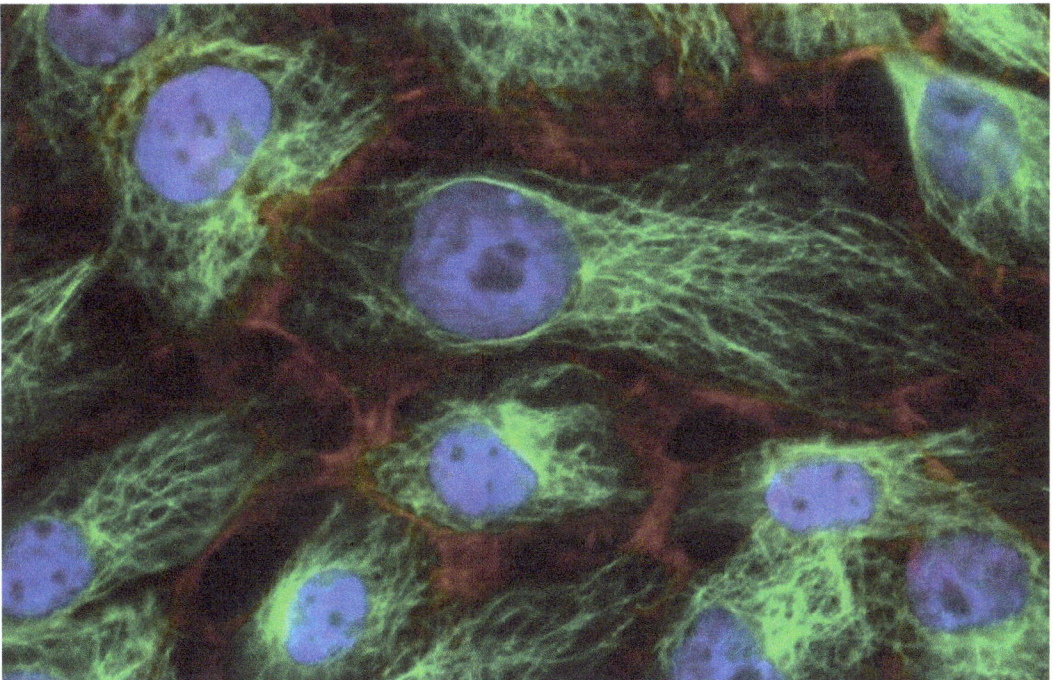

GENETIC MUTATIONS:

Cultures were created to expose the organism to various intensities of the mutagens and observed for a period of five years in order to analyze the effect of the mutagens on mutations in the organism. There were no observable mutations after fertilization and growth of the new unicellular organism. When exposed to high intensity of the mutagens, fertilization did not occur. Similarly, fertilized organisms that were exposed to low intensity of the mutagens did develop, but, with no observable mutations.

CONCLUSION:

The study of the effect of mutagens on paramecium did reveal that no significant mutations were observed. The only observed mutation was the change in color of the organism, which was due to the death of the unicellular organism, as a result of exposure to high intensity of the mutagens. It was observed, however, that an interesting organelle within the nucleus of unicellular organism was responsible for shutting down the whole system when exposed to high intensity of mutagens. This organelle, for now, is referred to as the MasterCodon. It seems to have an exerting influence on the whole organism and prevents any mutation that is lethal to the organism from taking place. Rather, than have a mutation taking place that could affect the metabolic activity in a negative manner, or, injurious to the organism, the MasterCodon reacts by completely shutting down the system, including all DNA and RNA genetic functioning. In other words, the MasterCodon seems to have a controlling effect on the essential genes and the non-essential genes.

Suggested Reading

Panno, Joseph (14 May 2014). The Cell: Evolution of the First Organism. Infobase Publishing. ISBN 9780816067367.

Bertrand, Jean-Claude; Caumette, Pierre; Lebaron, Philippe; Matheron, Robert; Normand, Philippe; Sime-Ngando, Télesphore (2015-01-26). Environmental Microbiology: Fundamentals and Applications: Microbial Ecology. Springer. ISBN 9789401791182.

Madigan, Michael T. (2012). Brock Biology of Microorganisms. Benjamin Cummings. ISBN 9780321649638.

Yaeger, Robert G. (1996). Protozoa: Structure, Classification, Growth, and Development. NCBI. ISBN 9780963117212. PMID 21413323. Retrieved 2018-03-23.

Yaeger, Robert G. (1996). Baron, Samuel (ed.). "Protozoa: Structure, Classification, Growth, and Development". University of Texas Medical Branch at Galveston. PMID 21413323. Retrieved 2020-07-07.

Goldfuß (1818). "Ueber die Classification der Zoophyten" [On the classification of zoophytes]. Isis, Oder, Encyclopädische Zeitung von Oken (in German). 2(6): 1008–1019. From p. 1008: *"Erste Klasse. Urthiere. Protozoa."* (First class. Primordial animals. Protozoa.) [Note: each column of each page of this journal is numbered; there are two columns per page.]

Scamardella JM (1999). "Not plants or animals: A brief history of the origin of Kingdoms Protozoa, Protista, and Protoctista" (PDF). International Microbiology. 2 (4): 207–221. PMID 10943416.

Ruggiero, Michael A.; Gordon, Dennis P.; Orrell, Thomas M.; Bailly, Nicolas; Bourgoin, Thierry; Brusca, Richard C.; Cavalier-Smith, Thomas; Guiry, Michael D.; Kirk, Paul M. (29 April 2015). "A Higher Level Classification of All Living Organisms". PLOS ONE. 10(4): e0119248. Bibcode:2015PLoSO..1019248R. doi:10.1371/journal.pone.0119248. PMC 4418965. PMID 25923521.

Cavalier-Smith, Thomas (1981). «Eukaryote kingdoms: seven or nine?». Bio Systems. 14 (3–4): 461–481. doi:10.1016/0303-2647(81)90050-2. PMID 7337818.

Cavalier-Smith, Thomas (December 1993). "Kingdom protozoa and its 18 phyla". Microbiological Reviews. **57**(4): 953–994. doi:10.1128/mmbr.57.4.953-994.1993. PMC 372943. PMID 8302218.

Cavalier-Smith, Thomas (23 June 2010). "Kingdoms Protozoa and Chromista and the eozoan root of the eukaryotic tree". Biology Letters. **6** (3): 342–345. doi:10.1098/rsbl.2009.0948. PMC 2880060. PMID 20031978.

Rothschild, Lynn J. (1989). "Protozoa, Protista, Protoctista: What's in a Name?". Journal of the History of Biology. **22** (2): 277–305. doi:10.1007/BF00139515. ISSN 0022-5010. JSTOR 4331095. PMID 11542176. S2CID 32462158.

Goldfuß, Georg August (1820). Handbuch der Zoologie. Erste Abtheilung [Handbook of Zoology. First Part.] (in German). Nürnberg, (Germany): Johann Leonhard Schrag. pp. XI–XIV.

Bailly, Anatole (1981-01-01). Abrégé du dictionnaire grec français. Paris: Hachette. ISBN 978-2010035289. OCLC 461974285.

Bailly, Anatole. "Greek-french dictionary online". www.tabularium.be. Retrieved 2018-10-05.

Hogg, John (1860). "On the distinctions of a plant and an animal, and on a fourth kingdom of nature". Edinburgh New Philosophical Journal. 2nd series. **12**: 216–225.

Scamardella, J. M. (December 1999). "Not plants or animals: a brief history of the origin of Kingdoms Protozoa, Protista and Protoctista". International Microbiology. **2** (4): 207–216. PMID 10943416.

Copeland, Herbert F. (September–October 1947). "Progress Report on Basic Classification". The American Naturalist. **81** (800): 340–361. doi:10.1086/281531. JSTOR 2458229. PMID 20267535.

Siebold (vol. 1); Stannius (vol. 2) (1848). Lehrbuch der vergleichenden Anatomie [Textbook of Comparative Anatomy] (in German). vol. 1: Wirbellose Thiere(Invertebrate animals). Berlin, (Germany): Veit & Co. p. 3.From p. 3: "Erste Hauptgruppe. Protozoa. Thiere, in welchen die verschiedenen Systeme der Organe nicht scharf ausgeschieden sind, und deren unregelmässige Form und einfache Organisation sich auf eine Zelle reduziren lassen." (First principal group. Protozoa.

Animals, in which the different systems of organs are not sharply separated, and whose irregular form and simple organization can be reduced to one cell.)

Dobell, C. (April 1951). "In memoriam Otto Bütschli (1848-1920) "architect of protozoology"". Isis; an International Review Devoted to the History of Science and Its Cultural Influences. 42 (127): 20–22. doi:10.1086/349230. PMID 14831973.

Taylor, F. J. R. 'Max' (11 January 2003). "The collapse of the two-kingdom system, the rise of protistology and the founding of the International Society for Evolutionary Protistology (ISEP)". International Journal of Systematic and Evolutionary Microbiology. 53(6): 1707–1714. doi:10.1099/ijs.0.02587-0. PMID 14657097

Whittaker, R. H. (10 January 1969). "New concepts of kingdoms or organisms. Evolutionary relations are better represented by new classifications than by the traditional two kingdoms". Science. 163 (3863): 150–160. Bibcode:1969Sci...163..150W. CiteSeerX 10.1.1.403.5430. doi:10.1126/science.163.3863.150. PMID 5762760.

Margulis, Lynn (1974). «Five-Kingdom Classification and the Origin and Evolution of Cells». In Dobzhansky, Theodosius; Hecht, Max K.; Steere, William C. (eds.). Evolutionary Biology. Springer. pp. 45–78. doi:10.1007/978-1-4615-6944-2_2. ISBN 978-1-4615-6946-6.

Cavalier-Smith, Thomas (August 1998). «A revised six-kingdom system of life». Biological Reviews. 73 (3): 203–266. doi:10.1111/j.1469-185X.1998.tb00030.x. PMID 9809012.

"New President's Address". protozoa.uga.edu. Retrieved 1 May 2015.

Singleton, Paul; Sainsbury, Diana (2001). Dictionary of microbiology and molecular biology. Wiley. ISBN 9780471941507.

Gooday, A.J.; Aranda da Silva, A. P.; Pawlowski, J. (1 December 2011). "Xenophyophores (Rhizaria, Foraminifera) from the Nazaré Canyon (Portuguese margin, NE Atlantic)". Deep-Sea Research Part II: Topical Studies in Oceanography. 58 (24–25): 2401–2419. Bibcode:2011DSRII..58.2401G. doi:10.1016/j.dsr2.2011.04.005.

Ghaffar, Abdul. "Blood and Tissue Protozoa". Microbiology and Immunology On-Line. Retrieved 2018-03-23.

Mylnikov, Alexander P.; Weber, Felix; Jürgens, Klaus; Wylezich, Claudia (August 2015). "Massisteria marina has a sister: Massisteria voersi sp. nov., a rare species isolated from coastal waters of the Baltic Sea". European Journal of Protistology. **51** (4): 299–310. doi:10.1016/j.ejop.2015.05.002. PMID 26163290.

Mitchell, Gary C.; Baker, J. H.; Sleigh, M. A. (1 May 1988). "Feeding of a freshwater flagellate, Bodo saltans, on diverse bacteria". The Journal of Protozoology. **35** (2): 219–222. doi:10.1111/j.1550-7408.1988.tb04327.x.

Ghaffar, Abdul. "Blood and tissue Protozoa". Microbiology and Immunology On-Line. Retrieved 2018-03-23.

"Trypanosoma brucei". parasite.org.au. Retrieved 2018-03-23.

"Microscopy of Entamoeba histolytica". msu.edu. Retrieved 2016-08-21.

Lehman, Don. "Diagnostic parasitology". University of Delaware. Retrieved 2018-03-23.

CHAPTER III:

PHYLUM–PORIFERA – "SOME ORDERLY EVOLUTIONARY TRENDS"

"One of the reasons I started taking this anti-evolutionary view, was ... it struck me that I had been working on this stuff for twenty years and there was not one thing I knew about it. That's quite a shock to learn that one can be so misled so long. ...so for the last few weeks I've tried putting a simple question to various people and groups of people. Question is: Can you tell me anything you know about evolution, any one thing that is true? I tried that question on the geology staff at the Field Museum of Natural History and the only answer I got was silence. I tried it on the members of the Evolutionary Morphology Seminar in the University of Chicago, a very prestigious body of evolutionists, and all I got there was silence for a long time and eventually one person said, 'I do know one thing — it ought not to be taught in high school'." Dr. Colin Patterson

The Phylum, Porifera, contains animal that are multicellular and heterotrophic organisms that lack cell walls, but produce sperm cells. They are different from other multicellular animals in that they do not have tissues and organs, and are referred to as sponges. Though, some of these organisms live in freshwater, nonetheless, majority of them inhabit deep seawater up to a depth of 5.5 miles. They are known to have been in existence for over 580 million years. For this study, the red boring sponge referred to as Cliona was examined (see Figure 3.1).

Figure 3.1: Picture of Blue Sponges.

Nutrition

Observations of Cliona in radioactive seawater did reveal that the organism gradually stopped feeding with high intensity of radioactive elements present. Similarly, when the organism was exposed gradually to increasing intensity of x-rays and ultraviolet rays, the organism completely stopped feeding. After a long period of time, the organism expired (see Figure 3.2).

Figure 3.2: Picture of Pink and Red Sponge.

Reproduction

Most sponges are hermaphrodites; however, there are some sponges that reproduced asexually. This asexual reproduction takes place by buds called gemmules which are aggregate of amoeboid cells called archeocytes. These cells do regenerate and give rise to other cells. Cliona species that are hermaphrodites reproduce through the use of sperm cell which originate from choanocytes and eggs cells that originates from archeocytes. The sperm and eggs are usually released in the water within the sponges' canals. The sperm cells are trapped by the choanocytes, which will carry this to a free floating egg within the parent atrium, or, to another sponge. The fertilized egg may be carried out by seawater and develop externally or, develop within a brood in the parent mesohyl. The developed embryo grows into a flagellated larva which swim out with the water current and after a while settles at the bottom of the sea, or, freshwater and develop into an adult sponge. The study did indicate that sponges do thrive in low intensity of ultra-violet rays. When the intensity was gradually increased, reproductive activities completely stopped. The organism, exposure to x-rays and radioactive chemicals created similar results.

Genetic Mutations

Cultures were created to expose the organism to various intensities of mutagens and observed for a period of five years in order to analyze the effect of mutagens on mutations in the organism. There was no observable mutation after fertilization and growth of the larva. When exposed to high intensity of the mutagens, fertilization did not occur. Similarly, fertilized zygotes that were exposed to low intensity of the mutagens did grow, but, with no observable mutations. Exposing the fertilized zygotes to high intensity of the mutagens resulted in the zygotes not thriving and not developing into the adult stage. Rather, the organism expired.

Conclusion

The study of the effect of mutagens on sponges did reveal that no significant mutations were observed. The only observed mutation was the change in color of the sponge, which was due to the death of the Cliona cells, as a result of exposure to high intensity of the mutagens. It was observed, however, that an interesting organelle within the nucleus of Cliona cells was responsible for shutting down the whole system when exposed to high intensity mutagens. This organelle, for now, is referred to as the MasterCodon. It seems to have an exerting influence on the whole organism and prevents any mutation that is lethal to the organism from taking place.

Rather, than have a mutation taking place that could affect the metabolic activity in a negative manner, or, injurious to the organism, the MasterCodon reacts by completely shutting down the system, including all DNA and RNA genetic functioning. In other words, the MasterCodon seems to have a controlling effect on the essential genes and the non-essential genes.

Suggested Reading

Bell JJ, McGrath E, Biggerstaff A, Bates T, Bennett H, Marlow J, Shaffer M. (2015) Sediment impacts on marine sponges. Marine Pollution Bulletin 94: 5–13. 10.1016/j.marpolbul.2015.03.030 [PubMed] [CrossRef] [Google Scholar]

Cárdenas P, Pérez T, Boury-Esnault N. (2012) Sponge systematics facing new challenges. In: Becerro MA, Uriz MJ, Maldonado M, Turon X. (Eds) Advances in Sponge Science: Phylogeny, Systematics, Ecology.Advances in Marine Biology 61: 79–209. 10.1016/B978-0-12-387787-1.00010-6 [CrossRef]

Dybowski W. (1880) Studien über die Spongien des russischen Reiches, mit besonderer Berücksichtigung der Spongien-Fauna des Baikal-Sees. Mémoires de l'Académie Impériale des sciences de St. Pétersbourg 7: 1–71. [Google Scholar]

Efremova SM. (2001) Sponges (Porifera). In: Timoshkin OA. (Ed.) Index of Animal Species Inhabiting Lake Baikal and its Catchment Area.Nauka, Novosibirsk, 182–192.

Efremova SM. (2004) New genus and new species of sponges from family Lubomirskiidae Rezvoj, 1936. In: Timoshkin OA. (Ed.) Index of Animal Species Inhabiting Lake Baikal and its Catchment Area.Nauka, Novosibirsk, 1261–1278.

Efremova SM, Goureeva MA. (1989) The problem of the origin and evolution of Baikalian sponges. The 1st Vereshchagin Baikal Conference, Irkutsk. Abstracts, 22–23

Hall TA. (1999) BioEdit: a user-friendly biological sequence alignment editor and analysis program for Windows 95/98/NT. Nucleic Acids Symposium Series 41: 95–98. [Google Scholar]

Harcet M, Bilandzija H, Bruvo-Madaric B, Cetkovic H. (2010) Taxonomic position of *Eunapius subterraneus* (Porifera, Spongillidae) inferred from molecular data – A revised classification needed? Molecular Phylogenetics and Evolution 54: 1021–1027. 10.1016/j.ympev.2009.12.019 [PubMed] [CrossRef]

Itskovich V, Gontcharov A, Masuda Y, Nohno T, Belikov S, Efremova S, Meixner M, Janussen D. (2008) Ribosomal ITS sequences allow resolution of freshwater sponge phylogeny with alignments guided by secondary structure prediction.

Journal of Molecular Evolution 67: 608–620. 10.1007/s00239-008-9158-5 [PubMed] [CrossRef] [Google Scholar]

Katoh K, Toh H. (2008) Recent developments in the MAFFT multiple sequence alignment program. Briefing on Bioinformatics 9: 286–298. 10.1093/bib/bbn013 [PubMed] [CrossRef] [Google Scholar]

Khalzov IA, Mekhanikova IV, Sitnikova TYa. (2018) First data on ectosymbiotic consortia of infusoria and prokaryotes associated with amphipods inhabiting the Frolikha underwater hydrothermal vent, Lake Baikal. Zoological Journal 97: 1525–1530. 10.1134/S0044513418120073 [CrossRef] [Google Scholar]

Khamidekh S. (1991) Analysis of anatomic and histological traits of sponges of Lubomirskiidae family. To the question of Baikal sponges taxonomy. PhD Thesis, Zoological Institute, Saint-Petersburg. [In Russian]

Khanaev IV, Kravtsova LS, Maikova OO, Bukshuk NA, Sakirko MV, Kulakova NV, Butina TV, Nebesnykh IA, Belikov SI. (2018) Current state of the sponge fauna (Porifera: Lubomirskiidae) of Lake Baikal: Sponge disease and the problem of conservation of diversity. Journal of Great Lakes Research 44: 77–85. 10.1016/j.jglr.2017.10.004 [CrossRef] [Google Scholar]

Kozhov MM. (1947) Animals of the Lake Baikal. Irkutsk regional Publishers, Irkutsk, 304 pp. [In Russian] [Google Scholar]

Kumar S, Stecher G, Tamura K. (2016) MEGA7: Molecular Evolutionary Genetics Analysis vertion 7.0 for bigger datasets. Molecular Biology and Evolution 33: 1870–1874. 10.1093/molbev/msw054 [PubMed] [CrossRef] [Google Scholar]

Lavrov DV. (2010) Rapid proliferation of repetitive palindromic elements in mtDNA of the endemic Baikalian sponge *Lubomirskia baicalensis*. Molecular Biology and Evolution 27: 757–760. 10.1093/molbev/msp317 [PubMed] [CrossRef] [Google Scholar]

Lavrov DV, Maikova OO, Pett W, Belikov SI. (2012) Small inverted repeats drive mitochondrial genome evolution in Lake Baikal sponges. Gene 505: 91–99. 10.1016/j.gene.2012.05.039 [PubMed] [CrossRef] [Google Scholar]

Maikova O, Khanaev I, Belikov S, Sherbakov D. (2015) Two hypotheses of the evolution of endemic sponges in Lake Baikal (Lubomirskiidae). Journal of

Zoological Systematics and Evolutionary Research 53: 175–179. 10.1111/.jzs.12086 [CrossRef] [Google Scholar]

Maikova O, Sherbakov D, Belikov S. (2016) The complete mitochondrial genome of *Baikalospongia intermedia* (Lubomirskiidae): description and phylogenetic analysis. Mitochondrial DNA. Part B: Resources 1: 569–570. 10.1080/23802359.2016.1172273 [CrossRef] [Google Scholar]

Maikova OO, Bukshuk NA, Itskovich VB, Khanaev IV, Nebesnykha IA, Onishchuk NA, Sherbakov DYu. (2017) Transformation of Baikal Sponges (Family Lubomirskiidae) after Penetration into the Angara River. Russian Journal of Genetics 53: 1343–1349. 10.1134/S1022795417120092 [CrossRef] [Google Scholar]

Maikova OO, Stepnova GN, Belikov SI. (2012) Variations in noncoding sequences of the mitochondrial DNA in sponges from family Lubomirskiidae. Doklady Biological Sciences 442: 46–48. 10.1134/S1607672912010140 [PubMed] [CrossRef] [Google Scholar]

Makushok ME. (1927a) Taxonomy of Baikal sponges. I. Genera *Lubomirskia* Dyb. and *Swartschewskia* n. nov. Russian Zoological Journal 7: 79–103. [In Russian] [Google Scholar]

Makushok ME. (1927b) Taxonomy of Baikal sponges. II. New genus of Baikal sponge fauna *Baicalolepis* nov. gen. and new species of the genus *Baicalolepis fungiformis* nov. sp. Russian Zoological Journal 7: 124–131. [In Russian] [Google Scholar]

Manconi R, Pronzato R. (2002) Suborder Spongillina subord. nov.: Freshwater sponges. In: Hooper JNA, Van Soest RWM. (Eds) Systema Porifera.A guide to the classification of sponges. Kluwer Academic/ Plenum Publishers, New York, Boston, Dordrecht, London, Moscow, 921–1020. 10.1007/978-1-4615-0747-5_97 [CrossRef]

Manconi R, Pronzato R. (2015) Phylum Porifera. In: Thorp J, Rogers DC. (Eds) Ecology and general biology: Thorp and Covich's freshwater invertebrates, 4th ed.Elsevier, Amsterdam, 133–157. 10.1016/B978-0-12-385026-3.00008-5 [CrossRef]

Manconi R, Pronzato R. (2019) Phylum Porifera. In: Rogers DC, Thorp J. (Eds) Keys to Palaearctic Fauna: Thorp and Covich's freshwater invertebrates, 4th ed.Elsevier, Amsterdam, 45–87. 10.1016/B978-0-12-385024-9.00003-4 [CrossRef]

Maniatis T, Fritsch EF, Sambrook J. (1984) Molecular Cloning. Mir, Moscow, 480 pp. [Russian Translation] [Google Scholar]

Masuda Y. (2009) Studies on the Taxonomy and Distribution of Freshwater Sponges in Lake Baikal. In: Müller WEG, Grachev MA. (Eds) Biosilica in Evolution, Morphogenesis, and Nanobiotechnology.Springer, Berlin/Heidelberg, 81–110. 10.1007/978-3-540-88552-8_4 [PubMed] [CrossRef]

Mats VD, Ufimtsev GF, Mandelbaum MM. (2001) The Baikal basin in the Cenozoic: Structure and geologic history. Novosibirsk, SBRAS Press, Branch "GEO", 252 pp. [In Russian] [Google Scholar]

Mazepova GF. (1995) General characteristics of Lake Baikal. In: Timoshkin OA, Melnik NG, Mazepova GF, Sheveleva NG. (Eds) Guide and key to pelagic animals of Baikal (with ecological notes).Nauka: Siberian Publishing Firm RAS, Novosibirsk, 23–24. [In Russian]

Pronzato R, Bavestrello G, Cerrano C. (1998) Morpho-functional adaptations of three species of *Spongia* (Porifera, Demospongiae) from a Mediterranean vertical cliff. Bulletin of Marine Science 63: 317–328. [Google Scholar]

Rezvoy PD. (1936) Freshwater sponges of the USSR. In: Rezvoy PD. (Ed.) The Fauna of the USSR.AS USSR, Moscow, 1–42.

Ronquist F, Huelsenbeck JP. (2003) MrBayes 3: Bayesian phylogenetic inference under mixed models. Bioinformatics 19: 1572–1574. 10.1093/bioinformatics/btg180 [PubMed] [CrossRef] [Google Scholar]

Rützler K. (1974) The burrowing sponges of Bermuda. Smithsonian contributions to zoology 165: 1–32. 10.5479/si.00810282.165 [CrossRef] [Google Scholar]

Sollas WJ. (1885) A Classification of the Sponges. Annals and Magazine of Natural History. 10.1080/00222938509459901 [CrossRef]

Swartschewsky BA. (1901) Short communication on Baikal sponge fauna. Mémoires de l'Académie Impériale des sciences de St. Pétersbourg 15: 1–7 [In Russian] [Google Scholar]

Swartschewsky BA. (1902) Materials on fauna of sponges of Lake Baikal. Memories of Kiev Society of Nature 17: 329–352. [In Russian] [Google Scholar]

Van Soest RWM, Boury-Esnault N, Hooper JNA, Rützler K, de Voogd NJ, Alvarez B, Hajdu E, Pisera AB, Manconi R, Schönberg C, Klautau M, Kelly M, Vacelet J, Dohrmann M, Díaz M-C, Cárdenas P, Carballo JL, Ríos P, Downey R, Morrow CC. (2019) . World Porifera Database. http://www.marinespecies.org/porifera

Veynberg E. (2009) Fossil Sponge Fauna in Lake Baikal Region. In: Müller WEG, Grachev MA. (Eds) Biosilica in Evolution, Morphogenesis, and Nanobiotechnology. Springer, Berlin Heidelberg, 185–205. 10.1007/978-3-540-88552-8_8 [PubMed] [CrossRef]

Weinberg E, Eckert C, Mehl D, Mueller J, Masuda Y, Efremova S. (1999) Extant and fossil spongiofauna from the underwater Academician ridge of Lake Baikal (SE Siberia). Memories of the Queensland Museum 44: 651–657. [Google Scholar]

Weinberg EV. (2005) Sponge fauna of Pliocene-Quaternary deposits of Baikal. PhD thesis, Saint-Petersburg: Zoological Institute. [In Russian]

Weltner W. (1895) Spongilliden studien III. Katalog und Verbreitung der bekannten Süsswasserschwämme. Archiv für Naturgeschichte 61: 114–144. [Google Scholar]

Werding B, Sanchez H. (1991) Life habits and functional morphology of the sediment infaunal sponges *Oceanapia oleracea* and *Oceanapia peltata* (Porifera, Haplosclerida). Zoomorphology 110: 203–208. 10.1007/BF01633004 [CrossRef] [Google Scholar]

CHAPTER IV:

PHYLUM—CNIDARIANS—"SOME LOST RELATIVES"

"There are only two possibilities as to how life arose. One is spontaneous generation arising to evolution; the other is a supernatural creative act of God. There is no third possibility. Spontaneous generation, that life arose from non-living matter was scientifically disproved 120 years ago by Louis Pasture and others. That leaves us with the only possible conclusion that life arose as a supernatural creative act of God. I will not accept that philosophically because I do not want to believe in God. Therefore, I choose to believe in that which I know is scientifically impossible; spontaneous generation arising to evolution." Dr. George Wald

The cnidarians are mostly animals like the jellyfish, sea anemones, corals and hydras and are generally radially symmetrical with over 11,000 species. At one end, the animals do have a mouth and tentacles, and most of them are free swimming. We chose the typical hydra, Obelia, belonging to the class hydrozoa for this study (see Figure 4.1).

Figure 4.1: Picture of Pink and Orange Jellyfish.

Body Structure and Movement

The hydras generally do have the polypoid body form which is cylindrical with the mouth and tentacles directed upward. The mouth opens into a gastrovascular cavity and does not have a posterior outlet. This cavity is lined internally by the gastroderm, and the outer layer of the hydra's body is covered by the epidermis. The middle layer between these two layers is called the mesoglea and generally does not contain cells. The hydras are mostly sessile animals, and movements are usually restricted to the extension and contraction of the body, and the to and fro movement of the tentacles. Observations were made to examine the effect of mutagens on structure and movement of Obelia. The animal was exposed to various levels of low, middle and high intensity of ultra-violet ray, x-rays, and radioactive carbon. During exposure to low level of these mutagens, Obelia continued with its normal movements, except that there were attempts to avoid the rays. When exposed to middle intensity, movement of Obelia ceased totally within the first three days of observations (see Figure 4.2).

Figure 4.2: Picture of a bunch of Orange Jellyfish.

Nutrition

Hydras generally feed on planktonic organisms and digestion is by secretion from the gland cells in the gastroderm. The partially digested food is absorbed through the engulfment of these food particles by the gastroderm cell and digestion is completed intracellularly. Obelia was observed under laboratory conditions to ascertain the effect of mutagens on nutrition. Under low intensity of x-rays, ultra-violet rays, and radioactive solutions, Obelia continued normal feeding over a long period of over six months with no changes in nutrition patterns. As the intensity of these mutagens was gradually increased, Obelia gradually decreased nutrition and finally became comatose under high intensity of the mutagens (see Figure 4.3).

Figure 4.3: Picture of Yellow and Pink Sea Anemones.

Reproduction

Obelia usually reproduces by asexual reproduction and the mode is by budding. The bud develops as a part of the hydroid colony, or, could float away and gets stuck on a substratum like rock, shell and other hard surface substances at the bottom of the sea. In a hydroid colony, one of the polyps becomes specialized as the reproductive polyp, and is called the gonozoid. It is from the tip of this gonozoid that the medusa buds are produced, and a new life cycle begins. The medusa buds do carry the germ cells in their gastroderm, and these eggs and sperm cells are deposited and fertilization takes place. The fertilized eggs develop into the planulae, which are free swimming larvae. These could become attached to the stolon of the original colony, or, dispersed away to form a new hydroid colony. From the observations, reproductive activities did increase dramatically when Obelia was exposed to low intensity levels of radioactive materials, and in the presence of low intensity of ultraviolet rays and x-rays. However, with a medium intensity of the mutagens, it was observed that reproductive activities decreased and there was disintegration of the colony. A sustained level of this intensity over a period of five years resulted in stunted growth, and less production of viable germ cells. Under high intensity of the mutagens, Obelia simply stopped reproduction, and the colonies withered away.

Genetic Mutations

Examination of isolated germ cells revealed that there were few mutations taking place. However, it was observed that as the level of the intensity of the mutagens was gradually increased, mutations occurring were not spontaneous enough to detect. The organism reactions to the mutagens were mainly to protect itself and most changes observed were peripheral. Exposure of Obelia to a long period, 5 years, of sustainable existence under the influence of the mutagens could only result in stunted growth, and color changes in adaptation to its changing environment. All the changes observed were phenotypical, and the genotype was not affected.

Conclusion

From the observations, the organism remained unchanged. The organism genome size of 3.5×10^7 remained unchanged. The effect of the organelle, the MasterCodon, was to consistently ensure that there was constancy to the metabolic activity of the organism and ensure that the genome size remains unchanged.

Suggested Reading

Arrigoni R, Kitano YF, Stolarski J, Hoeksema BW, Fukami H, Stefani F, Galli P, Montano S, Castoldi E, Benzoni F. 2014. A phylogeny reconstruction of the Dendrophylliidae (Cnidaria, Scleractinia) based on molecular and micromorphological criteria, and its ecological implications. *Zoologica Scripta* 43:661-688

Boekschoten GJ, Best M. 1988. Fossil and recent shallow water corals from the Atlantic Islands off Western Africa. *Zoologische Mededelingen* 62:99-112

Brito A, López C, Ocaña O, Herrera R, Moro L, Monterroso O, Rodríguez A, Clemente S, Sánchez JJ. 2017. Colonización y expansión en Canarias de dos corales potencialmente invasores introducidos por las plataformas petrolíferas. *Vieraea* 45:65-82

Cairns SD. 2001. A generic revision and phylogenetic analysis of the Dendrophylliidae (Cnidaria: Scleractinia) *Smithsonian Contributions to Zoology* 615:1-75

Cairns SD. 2009. *On line appendix: phylogenetic list of the 711 valid Recent azooxanthellate scleractinian species with their junior synonyms and depth ranges. Cold-water corals: the biology and geology of deep-sea coral habitats*. Cambridge: Cambridge University Press. 1-28

Cairns SD, Kitahara MV. 2012. An illustrated key to the genera and subgenera of the recent azooxanthellate Scleractinia (Cnidaria, Anthozoa), with an attached glossary. *ZooKeys* 227:1-47

Chen CA, Willis BL, Miller DJ. 1996. Systematic relationships between tropical Corallimorpharians (Cnidaria: Anthozoa: Corallimorpharia): utility of the 5.8s and internal transcribed spacer (ITS) regions of the rRNA transcription unit. *Bulletin of Marine Science* 59:196-208

Chevalier JP. 1966. Contribution á l'etude des Madréporaires des côtes occidentales de l'Afrique tropicale. *Bulletin de l'IFAN* 28(4):1356-1405

Creed JC, Fenner D, Sammarco P, Cairns S, Capel KCC, Junqueira AOR, Cruz I, Miranda RJ, Carlos-Junior L, Mantelatto MC+1 more. 2017. The invasion of the azooxanthellate coral *Tubastraea* (Scleractinia: Dendrophylliidae) throughout the world: history, pathways and vectors. *Biological Invasions* 19:283-305

Darriba D, Taboada GL, Doallo R, Posada D. 2015. jModelTest 2: more models, new heuristics and high- performance computing. *Nature Methods* 9:6-9

Fenner D. 2005. *Corals of Hawaii, a field guide to the hard, black and soft corals of Hawai'i and the Northwest Hawaiian Islands, including Midway*. Honolulu: Mutual Publishing.

Folmer O, Black M, Hoeh W, Lutz R, Vrijenhoek R. 1994. DNA primers for amplification of mitochondrial cytochrome c oxidase subunit I from diverse metazoan invertebrates. *Molecular Marine Biology and Biotechnology* 3:294-299

Friedlander AM, Ballesteros E, Fay M, Sala E. 2014. Marine communities on oil platforms in Gabon, West Africa: high biodiversity oases in a low biodiversity environment. *PLOS ONE* 9:e103709

Fukami H, Budd AF, Paulay G, Sole A, Chen CA, Iwao K, Knowlton N. 2004. Conventional taxonomy obscures deep divergence between Pacific and Atlantic corals. *Microbial Ecology* 427:0-3

Gélin P, Postaire B, Fauvelot C, Magalon H. 2017. Reevaluating species number, distribution and endemism of the coral genus *Pocillopora* Lamarck, 1816 using species delimitation methods and microsatellites. *Molecular Phylogenetics and Evolution* 109:430-446

Guindon S, Dufayard J-F, Lefort V, Anisimova M, Hordijk W, Gascuel O. 2010. New algorithms and mehtods to estimate maximum-likelihood phylogenies: Asessing the performance of PhyML 2.0. *Systematic Biology* 59:307-321

Hellberg ME. 2006. No variation and low synonymous substitution rates in coral mtDNA despite high nuclear variation. *Evolutionary Biology* 6:1-8

Hoeksema BW, Cairns S. 2018. World List of Scleractinia. Dendrophylliidae Gray, 1847. (accessed 07 January 2020)

Kearse M, Moir R, Wilson A, Stones-Havas S, Cheung M, Sturrock S, Buxton S, Cooper A, Markowitz S, Duran C+4 more. 2012. Geneious basic: an integrated and extendable desktop software platform for the organization and analysis of sequence data. *Bioinformatics* 28:1647-1649

Kitahara MV, Cairns SD, Stolarski J, Blair D, Miller DJ. 2010. A comprehensive phylogenetic analysis of the Scleractinia (Cnidaria, Anthozoa) based on mitochondrial CO1 sequence data. *PLOS ONE* 5:e11490

Kitahara MV, Fukami H, Benzoni F, Huang D. 2016. The new systematics of Scleractinia: Integrating molecular and morphological evidence. In: Goffredo S, Dubinsky Z, eds. *The cnidaria, past, present and future*. Springer. 41-59

Knapp ISS, Puritz JB, Bird CE, Whitney JL, Sudek M, Forsman ZH, Toonen RJ. 2016. ezRAD—an accessible next-generation RAD sequencing protocol suitable for non-model organisms_v3.2. Protocols.io. *Life Sciences Protocol Repository*

Kumar S, Stecher G, Tamura K. 2016. MEGA7: molecular evolutionary genetics analysis version 7.0 for bigger datasets. *Molecular Biology and Evolution* 33(7):1870-1874

Laborel J. 1974. West African reef corals an hypothesis on their origin. In: Proceedings of the Second International Coral Reef Symposium. 425-443

Lanfear R, Calcott B, Ho SYW, Guindon S. 2012. PartitionFinder: combined selection of partitioning schemes and substitution models for phylogenetic analyses. *Molecular Biology and Evolution* 29:1695-1701

Lin M, Luzon KS, Licuanan WY, Ablan-lagman MC, Chen CA. 2011. Seventy-four universal primers for characterizing the complete mitochondrial genomes of Scleractinian corals (Cnidaria; Anthozoa) *Zoological Studies* 50:513-524

López C, Clemente S, Almeida C, Brito A, Hernández M. 2015. A genetic approach to the origin of *Millepora* sp. in the eastern Atlantic. *Coral Reefs* 34(2):631-638

López C, Clemente S, Moreno S, Ocaña O, Herrera R, Moro L, Monterroso O, Rodríguez A, Brito A. 2019. Invasive *Tubastraea* spp. and Oculina patagonica and other introduced scleractinians corals in the Santa Cruz de Tenerife (Canary Islands) harbor: Ecology and potential risks. *Regional Studies in Marine Science* 29:100713

Milne-Edwards H, Haime J. 1848. Mémoire 3. Monographie des eupsammides. *Annales des Sciences Naturelles, Zoologie, Series 3* 10:65-114 pls. 5-9

Ocaña O, Hartog JCD, Brito A, Moro L, Herrera R, Martín J, Ramos A, Ballesteros E, Bacallado JJ. 2015. A survey on Anthozoa and its habitats along the Northwest African coast and some islands: new records, descriptions of new taxa and biogeographical, ecological and taxonomical comments. Part I. *Revista de la Academica Canaria de Ciencia* XXVII:9-66

Paz-García DA, García-de León FJ, Balart EF. 2015. Switch between Morphospecies of *Pocillopora* Corals. *The American Naturalist* 186:434-440

Ronquist F, Huelsenbeck JP. 2003. MrBayes 3: Bayesian phylogenetic inference under mixed models. *Bioinformatics* 19:1572-1574

Takabayashi M, Carter DA, Loh WKT, Hoegh-Guldberg O. 1998. A coral-specific primer for PCR amplification of the internal transcribed spacer region in ribosomal DNA. *Molecular Ecology* 7:925-931

Todd PA. 2008. Morphological plasticity in scleractinian corals. *Biological Reviews* 83:315-337

Toonen RJ, Puritz JB, Forsman ZH, Whitney JL, Fernandez-Silva I, Andrews KR, Bird CE. 2013. ezRAD: a simplified method for genomic genotyping in non-model organisms. *PeerJ* 1:e203

White TJ, Bruns T, Lee S, Taylor J. 1990. Amplification and direct sequencing of fungal ribosomal RNA genes for phylogenetics. In: Innis MA, Gelfand DH, Sninsky JJ, White TJ, eds. *PCR protocols. A guide to methods and application.* San Diego: Academic Press Inc.

Zibrowius H. 1973. Revision des espèces actuelles du genre *Enallopsammia* Michelotti, 1871, et description de E. marenzelleri, nouvelle espèces bathyle à large distribution: Ocean Indien et Atlantique Central (Madreporaria, Dendrophylliidae) *Beaufortia* 21:37-54

Zibrowius H. 1980. *Les Scléractiniaires de la Méditeranée et de I'Atlantique nordoriental,* Mémoires de I'Institut Océanographique. 11:284

CHAPTER V:

PHYLUM—PLATYHELMINTHES—"SOME WRIGGLY RELATIVES FOUND"

"Darwinian theory is the creation myth of our culture. It's the officially sponsored, government financed creation myth that the public is supposed to believe in, and that creates the evolutionary scientists as the priesthood… So we have the priesthood of naturalism, which has great cultural authority, and of course has to protect its mystery that gives it that authority—that's why they're so vicious towards critics." Phillip Johnson

The Platyhelminthes were first noticed to have come into existence in the Cambrian period around about 540 million years ago. They generally are free living marine and freshwater flatworms, including tapeworms and parasitic flukes. These animals are bilaterally symmetrical, and for observation, the human adult parasitic tapeworm, T.Saginatus was utilized. The flatworm belongs to the class Cestoda (see Figure 5.1).

Figure 5.1: Picture of a Tapeworm.

Body Structure

Most tapeworms are endoparasites living in the gut of vertebrate animal. Their bodies are segmented, tegument-like and are non-ciliated. The anterior end is the head, or, scolex with hooks, or suckers, which are adaptations to a parasitic life in gut. The scolex is attached to the other part of the body by a small, narrow neck. The main part of the body, called strobila, contains segmented parts called the proglottides. New proglottides are constantly being formed around the neck regions and the older proglottides are at the posterior ends. These segments are linear and flattened, and are generally long with continuous growth patterns. The study entails the use of isolated tapeworms in laboratory conditions that involved the initial exposure of the organism to low level radioactive solution, x-rays and ultraviolet rays. The organism was exposed independently to these mutagens and, also by exposure to two and three combination of the mutagen simultaneously. The organism continued its normal growth and the effect of the mutagens over a period of five years showed no visible effect on the tapeworm. Under medium intensity, it was observed that the tapeworm gradually become swollen with a lot of fluid absorption, then withered away and shriveled within three months. Under high intensity of the mutagens, the tapeworm did not survive up to one week.

Nutrition

The tapeworm feed generally by absorption of digested food material in the gut of its host. There is no organized digestive system in the tapeworm. Study of the tapeworm feeding was done in the laboratory away from the host. The tapeworm continued normal feeding when exposed to low radiation and low intensity of x-rays and ultraviolet rays. A medium dose of the mutagens resulted in the gradual cessation of feeding activities, and initial absorption of fluid leading to bloated teguments and finally, gradual cessation of all the feeding activities and the tapeworm shriveled and died off.

Reproduction

Tapeworm does have a complete reproductive system within each proglottid. There are gonophores on each proglottid, which is for copulation. Copulation generally takes place between two different proglottids belonging to the same tapeworm, or, if the host has more than one tapeworm, then, copulation could occur between two tapeworms. Tapeworms are hermaphroditic, and the fertilized eggs, which are enclosed in a shell, are continuously expelled through the gonophores into the host's intestine, where it is passed out through the anus with host's feces. The fertilized eggs could also be enclosed in a uterine sac within the posterior proglottids, and these teguments normally are detached from the main body and are expelled with the feces. The fertilized embryo, which develops into a round larva with hooks, do need an intermediate host, like cows, for further development. The larva, or, oncosphere, develops in the muscle of the cow into a cysticercoid, or, bladder worms with an inverted scolex and hooks. When human eats uncooked or, semi-cooked beef, the bladder worm develops into the adult, in the gut of the host. An attempt was made to simulate the growth of the tapeworm under laboratory conditions and by exposure to mutagens. With low doses of radioactive solution, the organism continued normal reproductive activities. Once exposed to medium intensity of x-rays and ultra-violet rays, the organism initially tried to escape from the mutagens by recoiling. Later, normal metabolic activity and reproductive activity gradually ceased. Under high intensity of the mutagens, the tapeworm did not survive past 36 hours. Exposure of the egg, the larva, and the bladder worm, individually, to a sustain level of medium intensity of mutagens for a period of one year, did not result in observable mutations. Most of the embryos did not survive. The few that survived did not exhibit any mutant variation (see Figure 5.2).

Figure 5.2: Picture of various Tapeworms.

Genetic Mutations

Isolated cells from the tapeworms in different stages of exposure to the mutagens were examined, and the chromosomes at these stages were observed. During this period of study, we did not observed any changes in chromosomes. There were no changes in gene number and size. No significant observable mutation took place. The genome size of the tapeworm remained constant at 5.2×10^7 during the duration of the study.

Conclusion

From the study of the flatworms, it was observed that the mutagens had very little effect on the organism in terms of creating mutants from the original species. Rather, a new organelle, called the MasterCodon, seems to exert an influence on the organism by resisting changes to DNA and chromosomal content of the organic cell. Rather than allowing the organism to mutate, the MasterCodon shut the whole metabolic system down and the organism stops living.

Suggested Reading

Morrone JJ, Crisci JV. Historical biogeography: introduction to methods. Annu. Rev. Ecol. Syst. 1995;26: 373–401.

Henderson IM. Biogeography without area? Austral Syst Bot. 1991;4: 59.

Cracraft J. Historical biogeography and patterns of differentiation within the South American avifauna: areas of endemism. Ornithol. Monogr. 1985;36: 49–84.

Crother BI, Murray CM. Ontology of areas of endemism. J Biogeogr. 2011;38(6): 1009–1015.

Morrone JJ. Biogeographical regionalisation of the Neotropical Region. Zootaxa. 2014;3782(1): 001–110.

DaSilva MB, Pinto-da-Rocha R, and Morrone JJ. Historical relationships of areas of endemism of the Brazilian Atlantic rain forest: a cladistic biogeographic analysis of harvestman taxa (Arachnida: Opiliones). Curr Zool. 2016;63(5): 525–535. pmid:29492012

Gámez N, Nihei SS, Scheinvar E, Morrone JJ. A temporally dynamic approach for cladistic biogeography and the processes underlying the biogeographic patterns of North American deserts. J Zoolog Syst Evol Res. 2014. https://doi.org/10.1111/jzs.12142.

Vane-Wright RI, Humphries CJ, Williams PH. 1991. What to protect? Systematics and the agony of choice. Biol Conserv. 1991;55: 235–254. https://doi.org/10.1016/0006-3207(91)90030-d

Rodrigues AS, Orestes Cerdeira J, Gaston KJ. Flexibility, efficiency, and accountability: Adapting reserve selection algorithms to more complex conservation problems. Ecography. 2000;23: 565–574. https://doi.org/10.1111/j.1600-0587.2000.tb00175.x

Díaz Gómez JM. Estimating ancestral ranges: Testing methods with a clade of Neotropical lizards (Iguania: Liolaemidae). PloS One. 2011;6: e26412. https://doi.org/10.1371/journal.pone.0026412 pmid:22028873

Martínez-Hernández F, Mendoza-Fernández AJ, Pérez-García FJ, Martínez-Nieto MI, Garrido-Becerra JA, Salmerón-Sánchez E, et al. (2015) Areas of endemism as a conservation criterion for Iberian gypsophilous flora: a multi-scale test using the NDM/VNDM program. Plant Biosyst. 2015;149(3): 483–493. https://doi.org/10.1080/11263504.2015.1040481

Rosen BR. From fossils to earth history: Applied historical biogeography. In: Myers A, Giller P, editors. Analytical biogeography: An integrated approach to the study of animal and plant distribution. Chapman and Hall, London; 1988. p. 437–481

Morrone JJ. On the identification of Areas of Endemism. Syst. Biol. 1994;43: 438–441.

Szumik C, Cuezzo F, Goloboff PA, Chalup A. An optimality criterion to determine areas of endemism. Syst Biol. 2002;51: 806–816. pmid:12396592

Szumik C, Goloboff PA. Areas of endemism: an improved optimality criterion. Syst Biol. 2004;53: 968–977. pmid:15764564

Porzecanski AL, Cracraft J. Cladistic analysis of distributions and endemism (CADE): Using raw distributions of birds to unravel the biogeography of the South American arid lands. J Biogeogr. 2005;32: 261–275. https://doi.org/10.1111/j.1365-2699.2004.01138.x

Dos Santos DA, Fernández HR, Cuezzo MG, Domínguez E. Sympatry inference and network analysis in biogeography. Syst. Biol. 2008;57: 432–448. pmid:18570037

Oliveira U, Brescovit AD, Santos AJ. Delimiting Areas of Endemism through Kernel Interpolation. PLoS ONE. 2015;10(1): e0116673. pmid:25611971

Hoffmeister CH, Ferrari A. Areas of endemism of arthropods in the Atlantic Forest (Brazil): an approach based on a metaconsensus criterion using endemicity analysis. Biol J Linn Soc. 2016;119(1): 126–144. https://doi.org/10.1111/bij.12802

Morales-Guerrero A, Miranda TP, Marques AC. Comparison between Parsimony Analysis of Endemicity (PAE), Endemicity Analysis (EA), and an alternative coding of Three-Distribution Statements based on hypothetical distributions. Syst Biodivers. 2017;15(5): 391–398. https://doy.org/10.1080/14772000.2016.1257519

Wheeler WC. Statistical Modeling of Distribution Patterns: A Markov Random Field Implementation and Its Application on Areas of Endemism. Syst. Biol. 2019;0(0): 1–15. https://doi.org/10.1093/sysbio/syz033

Müller P. The dispersal centres of terrestrial vertebrates in the Neotropical realm: a study in the evolution of the Neotropical biota and its native landscape. Biogeographica 2. 1973; 1–250.

Kinzey WG. 1982. Distribution of primates and forest refuges. In: Prance GT, Editor. Biological diversification in the tropics. Columbia University Press, New York, New York; 1982. p. 455–482.

Silva JMC, Casteleti CHM. Estado da biodiversidade da Mata Atlântica brasileira. In: Galindo–Leal C, Câmara IG, editors. Mata Atlântica: Biodiversidade, Ameaças e Perspectivas. 2005.

DaSilva MB, Pinto-da-Rocha R, DeSouza AM. A protocol for the delimitation of areas of endemism and the historical regionalization of the Brazilian Atlantic Rain Forest using harvestmen distribution data. Cladistics. 2015;31: 692–705.

Prado JR, Brennand PGG, Godoy LP, Libardi GS, Edson, Abreu F Jr, et al. Species richness and areas o endemism of oryzomyine rodents (Cricetidae, Sigmodontinae) in South America: an NDM/VNDM approach.s J Biogeogr. 2015;42: 540–551.

Platnick NI. On areas of endemism. Aust. Syst. Bot. 1991;4: 11–12.

CHAPTER VI:

PHYLUM - ASCHELMINTHES — "THE SLIMY RELATIVES ARE DISCOVERED!"

"A growing number of respectable scientists are defecting from the evolutionist camp ... moreover, for the most part these 'experts' have abandoned Darwinism, not on the basis of religious faith or biblical persuasions, but on scientific grounds, and in some instances, regretfully."
Wolfgang Smith

This Phylum is also referred to as Nematode which contains group of animals denoted as roundworms. The oldest nematodes are believed to have been in existence for 400 million years; and some scientists suggest that they could have been in existence for about one billion years. There are about 25,000 species of nematodes and they inhabit various environments. They are ubiquitously present everywhere from marine water to freshwater; from terrestrial environments like all soils in the tropics, mountains to Polar Regions; and from caves to all crevices. It has been estimated that there are about 4.4×10^{20} nematodes inhabiting the Earth's topsoil which is 60 billion for each human (see Figure 6.1).

Figure 6.1: Picture of Roundworms.

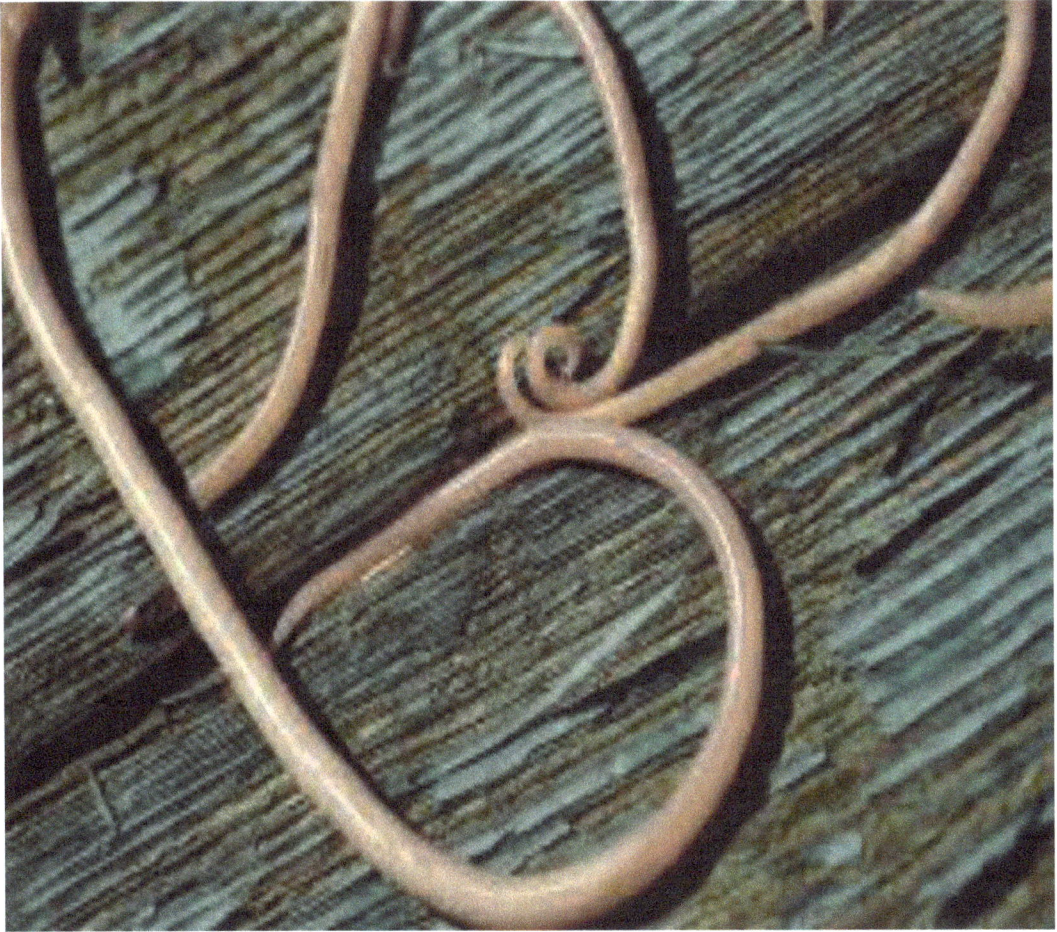

Nutrition and Excretion

Nematodes have oral cavities that are lined with cuticles and an anterior mouth with sharp stylets that are used to pierce into their prey for sucking. The oral cavity opens into a muscular, sucking pharynx and into the intestine that contains enzymes for digestion. The posterior part of the intestine is line by cuticles that forms the rectum and leads to the anus through which waste are disposed.

Reproduction

Few nematodes are hermaphrodites or androdioecious; however, most of these roundworms are dioecious (having separate male and female individuals). Both of the sexes possess one or two tubular gonads. In males, the sperms are produced at the posterior part of the gonad which travels through a muscular ejaculatory duct that is used to deposit sperm into the female vulva for fusion with the eggs. The fertilized eggs grow into the larvae that develop within the uterus and grow into young adults which are identical to the matured adult with no developed reproductive system (see Figure 6.2).

Figure 6.2: Picture of a Brown Worm.

Experiment with Mutagens

The roundworm, ascaris lumbricoides, was exposed to low intensity of radioactive solutions, and to ultraviolet rays and x-rays individually, and to two or three combination of the mutagens. Some of these cultures were exposed to the mutagens for about one year and some for five years. During this period of study, the nematode thrived, and occasionally exhibited avoidance reactions to the x-rays. There was no observed mutant variety from the newly produced zygotes, or the grown adults. Exposure of the nematodes to medium intensity of the same mutagens, and for cultures kept for six months, one year, and five years, produced phenotypic alterations in the colors and shrinkages in sizes and reduction in segmented cuticle. Generally the nematodes did not thrive and produced stunted larvae and most were aborted before the adult stages.

Again we did not observe any mutant varieties that were different from the original species. The same species were exposed to high intensity of the mutagens for culture of three months, six months, and one year. During this stage, the organism did not survive past the fifth day. The eggs that were left in the one year cultures and others did not develop. There was the re-introduction of some fertilized eggs from the one-year culture into normal environment, but the eggs did not develop further, which means that the germ cells have been destroyed by the mutagens.

Genetic Mutations

From the examination of all the cultures, an observation was made on the functioning of the nuclei in the cells of the nematode at different stages of exposure to the mutagens. It was revealed that there were virtually no changes in chromosomal content. Accidentally, we came across few cells that were exposed to medium intensity of the three mutagens used and found some with reduced chromosomal numbers, but the germ cells and zygote formed did not thrive. The genome of 8.2×10^7 bp, the gene size of 5.3 KB and gene number of 5,600 remained unchanged in all cultures, even in the mutant varieties with the reduced chromosomal numbers. However, it was observed that some nucleotides were displaced in the varieties exposed to medium intensity resulting in different codes for some proteins affecting the color of the organism. This same base shift resulted in the change in cuticle size, thickness and width of ring, including the ones with stunted growth. Again we observed the effect of the MasterCodon in controlling mutations.

Conclusion

It is apparent that there is a tendency to resist mutation by the organism. In other words the effect of mutagens on nematodes, though deleterious to the animal, is for the organism to resist mutations by dying, aborting, avoidance or total shutdown of its system. The desire to maintain its constant metabolic activity, the normal organic behavior and resistance to changes are so pre-eminent that the organism will rather die than mutate. It is for this reason that mutant varieties that escape this control do not survive. A discovery was made of an organelle called the MasterCodon which is the genetic control device that ensure that the organism maintains constancy from generation to generation, and prevents genotypic mutation from taking place. And, accidentally, when this mutation does take place, the MasterCodon shuts the animals system down completely, resulting in death of the organism.

Suggested Reading

Akaike H. (1974) A new look at the statistical model identification. IEEE transactions on automatic control 19: 716–723. 10.1109/TAC.1974.1100705 [CrossRef] [Google Scholar]

Bely AE, Wray GA. (2004) Molecular phylogeny of naidid worms (Annelida: Clitellata) based on cytochrome oxidase I. Molecular Phylogenetics and Evolution 30: 50–63. 10.1016/S1055-7903(03)00180-5 [PubMed] [CrossRef] [Google Scholar]

Cech G, Dózsa-Farkas K. (2005) Identification of *Fridericia schmelzi* sp.n. combining morphological characters and PCR-RFLP analysis. In: Pop VV, Pop AA. (Eds) Advances in Earthworm Taxonomy II.Cluj University Press, Cluj-Napoca, 99–118.

Cejka B. (1910) Die Oligochaeten der Russischen in den Jahren 1900–1903 unternommenen Nordpolarexpedition. I. Über eine neue Gattung der Enchytraeiden. Mémoires de l Académie Impériale des Sciences de St. Pétersbourg (Serie 8)29: 1–29.

Cejka B. (1912) Die Oligochaeten der Russischen in den Jahren 1900–1903 unternommenen Nordpolarexpedition. II. Über neue Bryodrilus- und Henlea-Arten. Mémoires de V Académie Impériale des Sciences de St. Pétersbourg (Serie 8)29: 1–19.

Cejka B. (1914) Die Oligochaeten der Russischen in den Jahren 1900–1903 unternommenen Nordpolarexpedition. III. Über neue Mesenchytraeus Arten. IV. Verzeichnis der während der Expedition gefundenen Oligochaeten-Arten. Mémoires de l 'Académie Impériale des Sciences de St. Pétersbourg (Serie 8)29: 1–32.

Chen J, Xie Z. (2015) *Mesenchytraeus asymmetriauritus*, a new enchytraeid (Annelida, Clitellata) from northeastern China. Proceedings of the Biological Society of Washington 128: 80–85. 10.2988/0006-324X-128.1.80

Christensen B, Dózsa-Farkas K. (1999) The enchytraeid fauna of the Palearctic tundra (Oligochaeta, Enchytraeidae). Biologiske Skrifter 52: 1–37. [Google Scholar]

Christensen B, Dózsa-Farkas K. (2006) Invasion of terrestrial enchytraeids into two postglacial tundras: north-eastern Greenland and the Arctic Archipelago of Canada (Enchytraeidae, Oligochaeta). Polar Biology 29: 454–466. 10.1007/s00300-005-0076-3 [CrossRef] [Google Scholar]

Coleman DC, Callaham JMA, Crossley JDA. (2018) Secondary production: activities of heterotrophic organisms – the soil fauna. Fundamentals of Soil Ecology. Academic Press, London, 47–76. 10.1016/B978-0-12-805251-8.00003-X [CrossRef]

Dózsa-Farkas K, Hong Y. (2010) Three new hemienchytraeus species (Enchytraeidae, Oligochaeta, Annelida) from Korea, with first records of other enchytraeids and terrestrial polychaetes (Annelida). Zootaxa 2406: 29–56. 10.11646/zootaxa.2406.1.2 [CrossRef] [Google Scholar]

Dózsa-Farkas K, Felföldi T, Hong Y. (2015) New enchytraeid species (Enchytraeidae, Oligochaeta) from Korea. Zootaxa 4006: 171–197. 10.11646/zootaxa.4006.1.9 [PubMed] [CrossRef] [Google Scholar]

Dózsa-Farkas K, Felföldi T, Nagy H, Hong Y. (2018) New enchytraeid species from Mount Hallasan (Jeju Island, Korea) (Enchytraeidae, Oligochaeta). Zootaxa 4496: 337–381. 10.11646/zootaxa.4496.1.27 [PubMed] [CrossRef] [Google Scholar]

Eisen G. (1905) Enchytræidæ of the west coast of North America: Harriman Alaska Expedition with cooperation of Washington Academy of Sciences. Doubleday, New York, 166 pp 10.5962/bhl.title.11673 [CrossRef] [Google Scholar]

Erséus C, Rota E. (2003) New findings and an overview of the oligochaetous Clitellata (Annelida) of the North Atlantic deep sea. Proceedings of the Biological Society of Washington 116: 892–900. https://www.biodiversitylibrary.org/page/34565774 [Google Scholar]

Erséus C. (2005) Phylogeny of oligochaetous Clitellata. Hydrobiologia 535: 357–372. 10.1007/s10750-004-4426-x

Erséus C, Rota E, Matamoros L, De Wit P. (2010) Molecular phylogeny of Enchytraeidae (Annelida, Clitellata). Molecular Phylogenetics and Evolution 57: 849–858. 10.1016/j.ympev.2010.07.005 [PubMed] [CrossRef] [Google Scholar]

Folmer O, Black M, Hoeh W, Lutz R, Vrijenhoek R. (1994) DNA primers for amplification of mitochondrial cytochrome c oxidase subunit I from diverse metazoan invertebrates. Molecular Marine Biology and Biotechnology 3: 294–9. [PubMed] [Google Scholar]

Ganin GN. (1984) Enchytraeids in deciduous forests of Middle Priamurye. Problemy Pochvennoy Zoologii, Proceedings of the VIII all-union meeting. Ashgabat, 68–69.

Hebert PDN, Ratnasingham S, DeWaard JR. (2003) Barcoding animal life: cytochrome c oxidase subunit 1 divergences among closely related species. Proceedings of the Royal Society B: Biological Sciences 270: S96–S99. 10.1098/rsbl.2003.0025 [PMC free article] [PubMed] [CrossRef]

Jamieson B, Tillier S, Tillier A, Justine J-L, Ling E, James S, Mcdonald K, Hugall A. (2002) Phylogeny of the Megascolecidae and Crassiclitellata (Annelida, Oligochaeta): combined versus partitioned analysis using nuclear (28S) and mitochondrial (12S, 16S) rDNA. Zoosystema 24: 707–734. [Google Scholar]

Krestov PV. (2003) Forest vegetation of easternmost Russia (Russian Far East). In: Kolbek J, Šrůtek M, Box EO. (Eds) Forest Vegetation of Northeast Asia.Springer, Dordrecht, 93–180. 10.1007/978-94-017-0143-3_5 [CrossRef]

Kumar S, Stecher G, Li M, Knyaz C, Tamura K. (2018) MEGA X: Molecular Evolutionary Genetics Analysis across computing platforms. Molecular Biology and Evolution 35: 1547–1549. 10.1093/molbev/msy096 [PMC free article] [PubMed] [CrossRef] [Google Scholar]

Lu Y, Xie Z, Zhang J. (2018) Preliminary taxonomical investigation of soil enchytraeids (Clitellata, Enchytraeidae) from south region of Tibet, China. Zootaxa 4496: 395–410. 10.11646/zootaxa.4496.1.29 [PubMed] [CrossRef] [Google Scholar]

Martinsson S, Erséus C. (2018) Cryptic diversity in supposedly Species-Poor genera of Enchytraeidae (Annelida: Clitellata). Zoological Journal of the Linnean Society. 184: 749–762. 10.1093/zoolinnean/zlx084 [CrossRef] [Google Scholar]

Martinsson S, Rota E, Erséus C. (2015a) Revision of *Cognettia* (Clitellata, Enchytraeidae): re-establishment of *Chamaedrilus* and description of cryptic species in the *sphagnetorum* complex. Systematics and Biodiversity 13: 257–277. 10.1080/14772000.2014.986555 [CrossRef] [Google Scholar]

Martinsson S, Rota E, Erséus C. (2015b) On the identity of *Chamaedrilus glandulosus* (Michaelsen, 1888) (Clitellata, Enchytraeidae), with the description of a new species. ZooKeys 501: 1–14. 10.3897/zookeys.501.9279

CHAPTER VII:

PHYLUM–MOLLUSCA – "OUR SHELLED COUSIN PAY A VISIT!"

"Hundreds of scientists who once taught their university students that the bottom line on origins had been figured out and settled are today confessing that they were completely wrong. They've discovered that their previous conclusions, once held so fervently, were based on very fragile evidences and suppositions which have since been refuted by new discoveries. This has necessitated a change in their basic philosophical position on origins. Others are admitting great weaknesses in evolution theory." Luther D Sutherland

The Phylum Mollusca is the second largest phylum of invertebrates' animals after the phylum Arthropoda. The phylum contains about 180,000 species of which nearly half have become extinct. This group of animals is believed to have been in existence for 540 million years in the Cambrian period (see Figure 7.1). The Snails were utilized for this study.

Figure 7.1: Picture of Snails in their habitats.

Respiration, Circulation & Excretion

In the land species, air currents move into the mantle cavity on the right side of the head and passes over the gills on the left sides before passing out. In the marine and fresh water species similar pattern except that water currents are being forced through the mantle. The gills are covered by large numbers of club-like projections which increase the surface area for enhanced air absorption. The organism does have a heart with a developed ventricle which receives blood from the gills. Additionally, snail does have an open vase system which allows air and other particles to be transported to all part of the body. The blood contains the respiratory pigment called hemocyte which allows air to be transported from the gills to the heart. The heart pumps blood which circulates to all other parts of the body by the pumping action of the ventricle through the aorta to all the tissue spaces of the body. Waste is collected from the coelom and passed through the kidneys or metanephridia. The cleansed blood is sent via gills back to the heart. The wastes are excreted as urine into the mantle cavity and flushed out of the cavity by circulating air or water currents.

Nutrition

The digestive system comprises of a mouth with a scraping or rasping organ called the radular. Supporting this anterior system is a cartilaginous skeleton called the odontophore, which enables the retractor muscle to project the rasping organ to the mouth for sucking and licking food particles into the mouth. Most mollusks are microphagous. But there are many that are carnivorous, herbivorous, scavengers, detritus feeders and parasitic feeders. Usually, once the food is brought into the mouth, mucus secrete by the salivary gland enveloped the food particles. This mass, then pass via the esophagus to the stomach or style sac, which is at the posterior part of the stomach with cilia for rotating the food particles around until they are broken into smaller fine particles. The acidic action of the enzyme breaks the mucous mass and the fine food particles which are conveyed by a duct to the digestive glands at the sides of the stomach. At this location, intracellular digestion takes place before it is absorbed into the coelom to the blood system. The undigested food particles are sent to the rear of intestines and sent out through the anus in the mantle cavity. Air or water currents force these waste out of the cavity.

Reproduction

Most mollusks do have separate sexes. There are many species that are also hermaphroditic. There are two gonads located at the anterior sides of the coelom. The eggs and sperms are normally released into the coelom where it is carried through the nephridia to the mantle cavity. External fertilization occurs when the eggs are deposited in string sacs and are fertilized by the sperms. A spiral cleavage occurs and zygote develops into the gastrula and then the trochophore larva. Anterior part of the larva has a ciliated mouth for feeding and movement. The trochophore normally develops into an advance larval stage called the liger. This then develop directly into the adult stage.

Experiment with Mutagens

The common snail was utilized in our experiment and exposed to low doses of radioactive elements, x-rays, and ultraviolet rays. The cultures contain different organisms that were exposed individually to one mutagen, and in combinations of two and three mutagens. The cultures were kept exposed to the mutagens for a period of three months, six months, one year, and five years. Initially, at low density of the mutagens, the snail continued normal feeding, normal metabolic activity, and reproduction patterns were unchanged. Later, it was observed that in all cultures with x-rays and radioactive elements present, that the snails started staying more within their shells after constant exposure for over six months. There were no detected changes in the organism, and the fertilized eggs did not produce any mutant variety (see Figure 7.2).

Figure 7.2: Picture of Snails.

Under medium intensity of the mutagens, it was observed that all the cultures exposed to x-rays and radioactive elements did not remain active beyond 30 days. Under the mutagens, it was observed that the organism attempted to escape, or avoid the mutagens as irritants. They then became erratic and defensive and in some cases became bloated. Apparently, the organisms were trying to maintain metabolic balances by absorbing more fluid to reduce the effect of the mutagens. Eventually, most metabolic activity ceased. The exposed zygotes and eggs did not flourish.

Attempts were made to fertilize eggs that were exposed to the mutagens over a period of five years. Most of the eggs have been actively destroyed and under-developed. The few that did develop exhibited stunted larval form and discoloration, but, none survived into the adult stage. Under high intensity of the mutagens, the organisms did not survive past ten days. There were no observed mutant varieties, because the organisms did not survive long enough.

Genetic Mutations

From the cultures exposed to low density x-rays and radioactive elements, we found few that had ring whorls enlarged, and few with smaller ring width. The surviving zygote exposed to medium intensity did exhibit mutant varieties in color changes and general reduction in size. Examination of nuclei from cells of the mutant varieties did, however, reveal that the chromosomal content of the snail remained unchanged. The gene size and number were unaffected. Mutations were restricted to shift in nucleotides bases that code slightly different proteins to allow the animal to adapt to changes in its environment, and these affected the color and sizes.

Conclusion

From this study, it was observed that the organism, always and constantly, resisted mutations. When few mutant varieties were observe, the zygotes, or the developed larva, did not survive. It was observed that another organelle closely associated with the chromosomes, which is labeled the MasterCodon, is the controlling nucleic material. The MasterCodon seems to exert influence on the organism by controlling the DNA, the genes and the chromosomes. The nature of this control is to maintain the constancy and uniqueness of the organism by resisting mutations, controlling the length of life of the organism, and ultimately shutting the whole system leading to the death of the organism.

Suggested Reading

Abbott, R. T. 1974. American seashells. The marine Mollusca of the Atlantic and Pacific coasts of North America. New York, Van Nostrand Reinhold Company. 663p. [Links]

Absalão, R. S. & de Paula, T. S. 2004. Shell morphometrics of three species of gadilid Scaphopoda (Mollusca) from the Southwestern Atlantic Ocean: comparing thediscriminating power of primary and secondary descriptors. Zootaxa 706:1-12. [Links]

Appolloni, M.; Smriglio, C.; Amati, B.; Lugliè, L.; Nofroni, I.; Tringali, L. P.; Mariottini, P. & Oliverio, M. 2018. Catalogue of the primary types of marine molluscan taxa described by Tommaso Allery Di Maria, Marquis of Monterosato, deposited in the Museo Civico di Zoologia, Roma. Zootaxa 4477(1):1-138. [Links]

Boissevain, M. 1906.The Scaphopoda of the Siboga Expedition, treated together with the known Indo-Pacific Scaphopoda. Uitkomsten op Zoologisch, Botanisch, Oceanographischen Geologisch Gebiedverzameld in Nederlandsch Oost-Indië 1899-1900 a anboord H.M. Sibogaonder commando van Luitenantter zee 1e. kl. G. F. Tydeman 54(Livraison 32):1-76, pls 1-6. [Links]

Caetano, C. H. S. & Absalão, R. S. 2005. A new species of the genus *Polyschides* Pilsbry & Sharp, 1898 (Mollusca, Scaphopoda, Gadilidae) from Brazilian waters. Zootaxa 871:1-10. [Links]

Caetano, C. H. S. & Santos, F. N. 2010. Mollusca, Scaphopoda, Gadilidae, *Striocadulus magdalensis* Gracia and Ardila, 2009: First record of the genus and species from Brazil. Check List 6(4):687-689. [Links]

Caetano, C. H. S. & Scarabino, V. 2009. Class Scaphopoda Bronn, 1862. *In*: Rios, E. C. ed. Compendium of Brazilian Sea Shells. Rio Grande, Evangraf, p. 444-457. [Links]

Caetano, C. H. S.; Garcia, N. & Lodeiros, C. J. M. 2007. First record of *Paradentalium infractum* (Odhner, 1931) (Mollusca, Scaphopoda, Dentaliidae) from the east coast of Venezuela. Brazilian Journal of Biology 67(4):797-798. [Links]

Caetano, C. H. S.; Scarabino, V. & Absalão, R. S. 2006. Scaphopoda (Mollusca) from the Brazilian continental shelf and upper slope (13° to 21°S) with descriptions of two new species of the genus *Cadulus* Philippi, 1844. Zootaxa 1267:1-47. [Links]

Caetano, C. H. S.; Scarabino, V. & Absalão, R. S.2010. Brazilian species of *Gadila* (Mollusca: Scaphopoda: Gadilidae): rediscovery of *Gadila elongata* comb. nov. and shell morphometrics. Zoologia 27(2):305-308.

Caprotti, E. 2015. *Antalis ariannae*, uma nuova specie mediterranea. Malacologia Mostra Mondiale 89:3-5. [Links]

Chistikov, S. D. 1983. [Modern molluscs of the family Entalinidae (Scaphopoda Gadilida), 4, subfamily Bathoxiphinae]. Zoologichesky Zhurnal 62(2):181-190. [originally in Russian]. [Links]

Dall, W. H. 1881. Reports on the results of dredging, under the supervision of Alexander Agassiz, in the Gulf of Mexico and in the Caribbean Sea (1877-78), by the United States Coast Survey Steamer "Blake", Lieutenant-Commander C.D. Sigsbee, U.S.N., and Commander J. R. Bartlett, U. S. N., commanding. XV. Preliminary report on the Mollusca. Bulletin of the Museum of Comparative Zoology 9(2):33-144. [Links]

Dall, W. H. 1889. Reports on the results of dredging, under the supervision of Alexander Agassiz, in the Gulf of Mexico (1877-78) and in the Caribbean Sea (1879-80), by the U. S. Coast Survey steamer "Blake", Lieut.-Commander C. D. Sigsbee, U. S. N., and Commander J. R. Bartlett, U. S. N., Commanding. XXIX-Report on the Mollusca. Part II. Gastropoda andScaphopoda. Bulletin of the Museum of Comparative Zoology 18:1-492. [Links]

Dall, W. H. 1890. Contributions to the Tertiary fauna of Florida, with especial reference to the Miocene silex-beds of Tampa and the Pliocene beds of the Caloosahatchie River. Part I. Pulmonate, opisthobranchiate and orthodont gastropods.Transactions of the Wagner Free Institute of Science of Philadelphia 3:1-200. [Links]

Dall, W. H. 1927. Small shells from dredgings off the southeast coast of the United States by the United States Fisheries Steamer "Albatross" in 1885 and 1886. Proceedings of the United States National Museum 70:1-134. [Links]

Dantas, R. J. S.; Laut, L. L. M. & Caetano, C. H. S. 2017. Diet of the amphi-Atlantic scaphopod *Fissidentalium candidum* in the deep waters of Campos Basin, south-eastern Brazil. Journal of Marine Biological Association of the United Kingdom 97(6):1259-1266. [Links]

Emerson, W. K. 1952. The scaphopod mollusks collected by the first Johnson-Smithsonian deep-sea expedition. Smithsonian Miscellaneous Collections 117(6):1-14. [Links]

Faller, S.; Rothe, B. H.; Todt, C.; Schmidt-Rhaesa, A. & Loesel, R. 2012. Comparative neuroanatomy of Caudofoveata, Solenosgastres, Polyplacophora, and Scaphopoda (Mollusca) and its phylogenetic implications. Zoomorphology 131:149-170. [Links]

Gofas, S.; Luque, A. A.; Templado, J. & Salas, C. 2017. A national checklist of marine Mollusca in Spanish waters. Scientia Marina 81(2):241-254. [Links]

Henderson, J. B. 1920. A monographof the East American scaphopod mollusks. United States National Museum Bulletin 111:1-177. [Links]

Jeffreys, J. G. 1877. New and peculiar Mollusca of the order Solenoconchia procured during the Valorous expedition. Annals and Magazine of Natural History 4(19):154-158. [Links]

Jeffreys, J. G.1882. On the Mollusca procured during the 'Lightning' and 'Porcupine' Expeditions, 1868-70 (Part V). Proceedings of the Zoological Society of London 1882:656-687.

Lamprell, K. L. & Healy, J. M. 1998. A revision of the Scaphopoda from Australian waters (Mollusca). Records of the Australian Museum Supplement 24:1-189. [Links]MolluscaBase. 2018a. Entalinidae Chistikov, 1979. Available at <Available at http://www.molluscabase.org/aphia.php?p=taxdetails&id=13695 >. Accessed on 20 April 2020. [Links]

MolluscaBase. 2018b. Scaphopoda. Available at <Available at http://molluscabase.org/aphia.php?p=taxdetails&id=104 >. Accessed on 20 April 2020. [Links]

Montserrat, F.; Guilhon, M.; Corrêa, P. V. F.; Bergo, N. M.; Signori, C. N.; Tura, P. M.; Maly, M. S.; Moura, D.; Jovane, L.; Pellizari, V.; Sumida, P. Y. G.; Brandini, F. P. & Turra, A. 2019. Deep-sea mining on the Rio Grande Rise (Southwestern

Atlantic): A review on environmental baseline, ecosystem services and potential impacts. Deep-Sea Research Part I, 145:31-58. [Links]

Nicklès, M. 1979. Scaphopodes de l'Ouest-Africain (Mollusqua, Scaphopoda). Bulletin du Muséum national d'Histoire naturelle, section A, Zoologie, Biologie et Écologieanimales (4ᵉserie) 1:41-77. [Links]

Pilsbry, H. A. & Sharp, B. 1897-1898. Scaphopoda. *In*: Tryon, G. W. ed. Manual of Conchology, 17. Philadelphia, Conchological Section, Academy of Natural Sciences. 280p. (1897: 1-144 p., 26 pls.; 1898: i-xxxii+145-280 p., pls.27-37). [Links]

Quinn, J. F. 1979. Biological results of the University of Miami deep-sea expeditions. 130. The systematics and zoogeography of the gastropod family Trochidae collected in the straits of Florida and its approaches. Malacologia 19(1):1-62. [Links]

Quinn, J. F. 1983. *Carenzia*, a new genus of Seguenziacea (Gastropoda: Prosobranchia) with the description of a new species. Proceedings of the Biological Society of Washington 96(3):355-364. [Links]

Reynolds, P. D. 2002. The Scaphopoda. Advances in Marine Biology 42:137-236. [Links]

Rios, E. C. 1994. Seashells of Brazil. 2ed. Rio Grande, Editora Fundação Universidade Rio Grande. 368p. [Links]

Rosenberg, G. 2014. A new critical estimate of named species-level diversity of the Recent Mollusca. American Malacological Bulletin 32(2):308-322. http://dx.doi.org/10.4003/006.032.0204 [Links]

Sahlmann, B. & van der Beek, J. 2016. A new scaphopod, *Dentalium humboldti* n. sp., from the Concepción Methane Seep off Chile (Mollusca: Scaphopoda). SchriftenzurMalakozoologie 29:41-48. [Links]

Sahlmann, B. & Wiese, V. 2016. A new scaphopod from Yemenite waters, *Tesseracme arabica* n. sp. (Mollusca: Scaphopoda). SchriftenzurMalakozoologie 29:33-40. [Links]

Sahlmann, B.; van der Beek, J. & Wiese, V. 2016. *Fissidentalium (Compressidentalium) pseudohungerfordi* n. sp., a well known undescribed scaphopod in the group of *Fissidentalium* (*Compressidentalium*) *hungerfordi* (Pilsbry& Sharp 1897) (Mollusca: Scaphopoda). SchriftenzurMalakozoologie 29:19-32. [Links]

Salvador, R. B.; Cavallari, D. C.& Simone, L. R. 2014. Seguenziidae (Gastropoda: Vetigastropoda) from SE Brazil collected by the Marion Dufresne (MD55) expedition. Zootaxa 3878(6):536-550. [Links]

Scarabino, F. 2003. Lista sistemática de los Aplacophora, Polyplacophora y Scaphopoda de Uruguay. Comunicaciones de la Sociedad Malacológica del Uruguay 8(78-79):191-196. [Links]

Scarabino, V. 1970. Las espécies del genero *Cadulus* Philippi, 1844 (Moll. Scaphopoda) em el Atlantico Sudoccidental (Lat. 24°S a 38°S). Comunicaciones de la Sociedad Malacológica del Uruguay 3(19):39-48.

Scarabino, V. 1975. Class Scaphopoda. *In*: Rios, E. C. ed. Brazilian marine mollusks iconography. Rio Grande, Museu Oceanográfico do Rio Grande, p. 180-186. [Links]

Scarabino, V. 1985. Class Scaphopoda Bronn, 1862. *In*: Rios, E. C. ed. Seashells of Brazil. Rio Grande, Museu Oceanográfico do Rio Grande , p. 196-202. [Links]

Scarabino, V. 1986a. Nuevos taxa abisales de la clase Scaphopoda (Mollusca). Comunicaciones Zoologicas del Museo de Historia Natural de Montevideo 11(155):2-19. [Links]

Scarabino, V. 1986b. Systematics of Scaphopoda, I. Three new bathyal and abyssal taxa of the order Gadilida from South and North Atlantic ocean. Comunicaciones Zoológicas del Museo de Historia Natural de Montevideo 11(161):1-15. [Links]

Scarabino, V. 1994. Class Scaphopoda Bronn, 1862. *In*: Rios, E. C. ed. Seashells of Brazil. 2ed. Rio Grande, FURG, p. 305-310. [Links]

Scarabino, V. 1995. Scaphopoda of the tropical Pacific and Indian waters,with descriptions of 3 new genera and 42 new species. *In*: Bouchet, P. ed. Résultats des Campagnes Musorstom, v.14. Mémoires du Muséum national d'Histoire naturelle 167:189-379. [Links]

Scarabino, V. 2008. New species and new records of scaphopods from New Caledonia. *In*: Héros, V.; Cowie, R. H. & Bouchet, P. eds. Tropical Deep-Sea

Benthos 25.Paris, Mémoires du Muséum national d'Histoire naturelle 196:215-268. [Links]

Scarabino, V. & Caetano, C. H. S. 2008. On the genus *Heteroschismoides* Ludbrook, 1960 (Scaphopoda: Gadilida: Entalinidae), with descriptions of two new species. The Nautilus 122(3):171-177. [Links]

Scarabino, V.; Caetano, C. H. S. & Carranza, A. 2011. Three new species of the deep-water genus *Bathycadulus* (Mollusca, Scaphopoda, Gadilidae). Zootaxa 3096:59-63. [Links]

Scarabino, V. & Scarabino, F. 2010. A new genus and thirteen new species of Scaphopoda (Mollusca) from the tropical Pacific Ocean. Zoosystema 32(3):409-423. [Links]

Scarabino, V. & Scarabino, F. 2011. Ten new bathyal and abyssal species of Scaphopoda from the Atlantic Ocean. The Nautilus 125(3):127-136. [Links]

Sigwart, J. D.; Sumner-Rooney, L. H.; Dickey, J. & Carey, N. 2017. The scaphopod foot is ventral: more evidence from the anatomy of *Rhabdus rectius* (Carpenter, 1864) (Dentaliida: Rhabdidae). Molluscan Research 37(2):79-87. [Links]

Silva-Filho, G. F. S.; Pinto, S. L. & Alves, M. S. 2010. Two new speciesof the genus *Gadila* Gray, 1847 (Mollusca, Scaphopoda, Gadilidae) from Brazilian coast. Revista Nordestina de Zoologia 4(1):48-53. [Links]

Silva-Filho, G. F. S.; Tenório, D. O.; Pinto, S. L. & Alves, M. S. 2012. Mollusca Scaphopoda Bronn, 1862 da Costa Nordeste do Brasil. Tropical Oceanography 40(1):29-103. [Links]

Silva-Filho, G. F. S.; Tenório, D. O.; Pinto, S. L. & Alves, M. S. 2016. Scaphopoda (Mollusca) of Fernando de Noronha–Pernambuco and Atol das Rocas–Rio Grande do Norte, Brazil. Revista Nordestina de Zoologia 10(2):52-73.

Simone, L. R. L. 2009. Comparative morphology among representatives of main taxa of Scaphopoda and basal protobranch Bivalvia (Mollusca). Papéis Avulsos de Zoologia 49(32):405-457. [Links]

Smith, A. M. & Spencer, H. G. 2016. Skeletal mineralogy of scaphopods: an unusual uniformity. Journal of Molluscan Studies 82(2):344-348. [Links]

Souza, L. S.; Araújo, I. C. V. & Caetano, C. H. S. 2013. A commented list of Scaphopoda (Mollusca) found along the Brazilian coast, with two new synonymies in the genus *Gadila* Gray, 1847. Biota Neotropica 13(2):228-235. [Links]

Steiner, G. & Dreyer, H. 2003. Molecular phylogeny of Scaphopoda (Mollusca) inferred from 18S rDNA sequences: support for a Scaphopoda-Cephalopoda clade. Zoologica Scripta 32(4):343-356. [Links]

Steiner, G. & Kabat, A. R. 2001. Catalogue of supraspecific taxa of Scaphopoda (Mollusca). Zoosystema 23(3):433-460. [Links]

Steiner, G. & Kabat, A. R. 2004. Catalog of species-group names of Recent and fossil Scaphopoda (Mollusca). Zoosystema 26(4):549-726. [Links]

Sumner-Rooney, L. H.; Schrödl, M.; Lodde-Bensch, E.; Lindberg, D. R.; Heß, M.; Brennan, G. P. & Sigwart, J. D. 2015. A neurophylogenetic approach provides new insight to the evolution of Scaphopoda. Evolution & Development 17(6):337-346. [Links]

Townsend, C. H. 1901. Dredging and other records of the United States Fish Commission Steamer Albatross, with bibliography relative to the work of the vessel. United States Fish Commission Report for 1900 1:387-562.[Links]

Verrill, A. E. 1885. Third catalogue of Mollusca recently added to the fauna of the New England coast and the adjacent parts of the Atlantic, consisting mostly of deep-sea species, with notes on others previously recorded. Transactions of the Connecticut Academy of Arts and Sciences 6:395-452. [Links]

Vilela, P. M. S.; Souza, L. S. & Caetano, C. H. S. 2019. Larval and early post-larval shell of three deep-sea Scaphopoda (Mollusca) from the southwest Atlantic. Molluscan Research 39(1):35-43. [Links]

Warén, A. 1980. Marine Mollusca described by John Gwyn Jeffreys, with the location of the type material. Conchological Society of Great Britain and Ireland Special Publication 1:1-60.

Warén, A. 1991. New and little known Mollusca from Iceland and Scandinavia. Sarsia 76:53-124. [Links]

Watson, R. B. 1886. Report on the Scaphopoda and Gasteropoda collected by H.M.S. Challenger during the years 1873-1876. Report on the Scientific Results of the Voyage of H.M.S. Challenger, Zoology 15(2):1-680, 692-756. [Links]

Zamudio, K. R.; Kellner, A.; Serejo, C.; Britto, M. R.; Castro, C. B.; Buckup, P. A.; Pires, D. O.; Couri, M.; Kury, A. B.; Cardoso, I. A.; Monné, M. L.; Pombal Jr., J.; Patiu, C. M.; Padula, V.; Pimenta, A.D.; Ventura, C. R. R.; Hajdu, E.; Zanol, J.; Bruna, E. M.; Fitzpatrick, J. & Rocha, L.A. 2018. Lack of Science support fails Brazil. Science 361(6409):1322-1323.

CHAPTER VIII:

PHYLUM-ANNELIDA "OUR CHITINOUS COUSINS PAY A VISIT"

"The fact that a theory so vague, so insufficiently verifiable, and so far from the criteria otherwise applied in 'hard' science has become a dogma can only be explained on sociological grounds." (Ludwig von Bertalanffy, biologist) 'I had motive for not wanting the world to have a meaning; consequently assumed that it had none, and was able without any difficulty to find satisfying reasons for this assumption. The philosopher who finds no meaning in the world is not concerned exclusively with a problem in pure metaphysics, he is also concerned to prove that there is no valid reason why he personally should not do as he wants to do, or why his friends should not seize political power and govern in the way that they find most advantageous to themselves. ... For myself, the philosophy of meaninglessness was essentially an instrument of liberation, sexual and political." Aldous Huxley

The animals belonging to the phylum Annelida were thought to have evolved around 535 million years ago in the Cambrian period. Most of these animals are the Nereids, Earthworms, and the Leeches. Most of these animals are to be found in marine and fresh waters. For this study, the earthworm, lumbricus, was chosen (see Figure 8.1).

Figure 8.1: Picture of an Earthworm.

Body Structure and Movement

The body of the earthworm is divided into similar parts, or segments that are in a linear series from the anterior to the posterior axis. The anterior end carries the head, or prostomium and it is not segmented. The posterior end of the body, called, the pygidium carries the anus. The segments are separated from each other by a septum. The segments are produced near the pygidium and use muscle contraction to move the organism along in a crawling manner. The segments have small bristles called setae to anchor to the substratum as they move along.

Nutrition

The earthworms are deposit feeders and do feed on decomposing plant matter in the soil. The decomposed soil matter circulates within their guts. The undigested soil casings are normally left on the surface and which have been found to be beneficial to farming. The mouth opens into the muscular pharynx, which could be everted as pump for ingestion of food. The food is moved down to the esophagus which have modifications that include the crop and the gizzards. The crop act as storage for food and the gizzards help to grind the food. There are also calciferous glands on the dorsal wall of the esophagus that excrete excess calcium taken in with the food, which are passed into the gut and, are expelled with other waste through the anus. The gut is surrounded by chloragogen cells that enable the synthesis and storage of glycogen and fat, the de-amination of amino acids, and the synthesis of urea.

Circulation, Respiration and Excretion

Earthworms have a closed blood vascular system. There is a contractive and longitudinal vessel above the gut, which function as the heart, and pumps blood anteriorly. There are paired vessels around the gut, which carry blood to a ventral longitudinal vessel, in which blood flows posteriorly. The ventral longitudinal blood vessel carry blood to all the body segments; and from each segment, blood is moved by a vessel to the gut, and through paired vessels to other structures in the segment. The blood does have hemoglobin for transportation of gases. Gas exchange is generally by absorption through the external wall of the body directly into the vascular system. The primary excretory organ is the metanephridia. And, in each segment there is a pair of nephridia. The primary function of this excretory organ is mainly that of, water balance, and to maintain a balance in the coelom.

Reproduction

The earthworms are hermaphrodites. Both the male and female gonads are present and restricted to a few segments at the anterior third of the body. Some six to eight of these segments have been fused externally together to form a region called the clitellum. This area is responsible for secreting fluids from glands in this region, which are used for the reproductive processes. The eggs generally mature in the coelomic fluid of this region, and are passed out through paired oviducts. About two or three segments do have paired testes, and leads to the pouch like seminal vesicles. And from here, there are paired sperm ducts which pass through about five segments before opening to the exterior on the ventral surface. Copulation takes place when two earthworms come together and fuses around the clitellum area with their anterior ends facing opposite sides. They will exchange sperms and this process could last for up to three hours. The sperms deposited by each worm are transferred to the seminal receptacles of the corresponding worm. During this process, secretion from glands in the clitellum is responsible for adherence of the worms. More secretions from this region are, also, responsible for the production of cocoons. Food source in the form of albumin and egg yolk are deposited in the cocoon with the fertilized eggs. The cocoon, after collecting eggs from the gonophore and sperm from the seminal receptacles, slips off at the anterior end and is left at the bottom of the mud, or, in the soil. Generally, only one egg will develop in the cocoon and feed on the deposited food within the cocoon.

Fertilization takes place in the cocoon and develops directly into the young adults or worms (see Figure 8.2).

Figure 8.2: Picture of an Earthworm.

Experiment with Mutagens

The earthworms were exposed in cultures with low density of radioactive elements, x-rays and ultraviolet rays. The organism was exposed to one mutagen at a time, and, then, in combination of two and three mutagens. Some of the cultures were exposed to the mutagens for periods of six months, one year, and five years. The adult organism, germ cells and fertilized zygotes were also exposed to the mutagens within the same time frame. The culture exposed to low intensity of the ultraviolet rays continued normal metabolic activities. The cultures exposed to low intensity of the x-rays and radioactive elements, initially thrived but did not survive past the ninth month. The germ cells that were exposed to the x-rays and radioactive elements for three, and five years periods were then allowed to be fertilized. Most of the eggs did not develop. From the cultures, it was observed that few mutant varieties exhibited discoloration, shortened bodies, fewer segments, and underdeveloped internal organs. All of the zygotes did not survive past the second week of adult stages.

Exposure of similar cultures to medium intensity of the mutagens resulted in the organisms remaining dormant. The cultures with the ultraviolet rays did not survive past the first week. Initially, it was observed that the earthworm became swollen which could be attributed to the function of the metanephridia in its attempt to absorb more fluid and reduce the effect of the mutagens.

The exposure of the earthworm to high density of the mutagens did not produce observable results since the animal did not survive. The germ cells exposed to these high doses did not thrive.

Genetic Mutations

Most of the cultures in the experiment did not produce any significant mutations. The few observed mutant varieties were only different in color and body segments; and they did not survive. Some of the organisms did have undeveloped sensory, excretory and reproductive organs. An atrophied gut was observed in one of the mutant varieties. All the organisms did not survive up to seven days. Cells from the mutant varieties, and from normal earthworms, were examined to ascertain any effect on the nuclei. It was observed that the chromosomes were unchanged in all the cultures. The genome of 8.7×10^7 bp remained unchanged in all these cultures, including the few mutant varieties. The production of different colors and size differentiations in the earthworms exposed to the low doses of the mutagens, and the observed differences in the mutant varieties, were activities that were controlled by an organelle associated with the chromosomes termed the MasterCodon by the author.

Conclusion

The study of the earthworm did reveal that the organism seems to resist any attempt to mutate. Mutation that did occur was only meant to adapt the organism to its changing environment. The mutations were not significant enough to change the total characteristics and organs of the organism. Few mutations that were observed were phenotypical and not genotypical. The organelle that seems to be exerting this influence is the MasterCodon. This organelle ensures that the organism maintain its constancy and uniqueness. Any attempt to make the change permanent resulted in the shutdown of the organism's system. In other words, the MasterCodon seems to control the organism's life, the life span, metabolic activities, health and eventually death. Further, the MasterCodon in the nuclei of the organism's cells seem coded to control the action of the DNA in protein synthesis in such a way as to ensure that the uniqueness of the organism is preserved.

Suggested Reading

Al-Abbad M. Y.M. New records of twelve species of Oligochaeta (Naididae and Aeolosomatidae) from the Southern Iraqi Marshes. Iraq. Jordan Journal of Biological Sciences. 2012;5(2):105–112. [Google Scholar]

Balik S., Ustaoğlu M. R., Yildiz S. Oligochaeta and Aphanoneura (Annelida) fauna of the Gediz Delta (Menemen-Izmir) Turk J. Zool. 2004;28:183–197. [Google Scholar]

Beddard F. E. On the geographical distribution of earthworms. Proc. Zool Soc London. 1894;1983:733–738. [Google Scholar]

Beddard F. E. Annelida. Oligochaeta. Kew Bull Misc Info Addit Ser. 1906;5:66–67. [Google Scholar]

Benham W. B. A new English genus of aquatic Oligochaeta (*Sparganophilus*) belonging to the family Rhinodrilidae. Q. J. Microsc. Soc. 1892;34:155–179. [Google Scholar]

Berger C. J., Füreder L., Salvenmoser W. Aquatic microcosm: Stone crayfish as a potential arena for a symbionts network; Proceedings of the IAA 19th Conference; Innsbruck, Austria. 2012.2012. [Google Scholar]

Bird G. J. Distribution, life cycle and population dynamics of the aquatic enchytraeid *Propappus volki* (Oligochaeta) in an English chalkstream. http://dx.doi.org/10.1111/j.1600-0587.1982.tb01019.x. Ecography. 1982;5(1):67–75. doi: 10.1111/j.1600-0587.1982.tb01019.x. [CrossRef] [Google Scholar]

Blakemore R. J. Whither Octochaetidae? – A review of its family status (Annelida: Oligochaeta) In: Pop V. V., Pop A. A., editors. Advances in earthworm taxonomy II. Cluj Univ. Press; Cluj: 2005. 63-84 [Google Scholar]

Blakemore R. J. Hibernian reports of a new Franco-Iberian worm (Oligochaeta: Megadrilacea: Lumbricidae) Opuscula Zoologica Budapest. 2012;43(2):121–130.

Blakemore R. J. Japanese earthworms revisited a decade on (Oligochaeta: Megadrilacea) Zoology in the Middle East. 2012;58(4):15–22. doi: 10.1080/09397140.2012.10648981. [CrossRef] [Google Scholar]

Bleidorn Christoph, Vogt Lars, Bartolomaeus Thomas. New insights into polychaete phylogeny (Annelida) inferred from 18S rDNA sequences. http://dx.doi.org/10.1016/s1055-7903(03)00107-6. Molecular Phylogenetics and Evolution. 2003;29(2):279–288. doi: 10.1016/s1055-7903(03)00107-6. [PubMed] [CrossRef] [Google Scholar]

Bouché M. B. Remarques sur quelques Lumbricina de France et consequences de la decouverte des nouveaux taxons Vignysinae (Subfam. nov.) et Diporodrilidae (Fam. nov.) Pedobiologia. 1970;10:246–256. [Google Scholar]

Bouché M. B. Lombriciens de France. Ecologie et systématique. Annales de Zoologie-Ecologie Animale. 1972;numero hors-série:1–671. [Google Scholar]

Bouché M. B. The establishment of earthworm communities. In: Satchell J. E., editor. Earthworm Ecology: From Darwin to Vermiculture. Chapman and Hall; London: 1983. [CrossRef][Google Scholar]

Bouché M. B., Qiu J-P. Un nouveau *Sparganophilus* (Annelida: Oligochaeta) d'Europe, avec considérations paléogéographiques sur les Lumbricina. Doc. Pédozool. Intégrol. 1998;16:178–180. [Google Scholar]

Branko J., Bilandžija H., Cukrov M. Distribution of the Dinaric cave-dwelling tube worm *Marifugia cavatica* Absolon & Hrabe, 1930 in Croatia. Poster presented at 21st Int. Conference on Subterranean Biology, Abstract book, Košice, Slovačka, 02.-07.09.2012 2012

Brinkhurst R. O. Evolutionary relationships within the Clitellata: an update. Megadrilogica. 1994;5:109–112. [Google Scholar]

Briones María Jesús Iglesias Briones, Morán Paloma, Posada David. Are the sexual, somatic and genetic characters enough to solve nomenclatural problems in lumbricid taxonomy? http://dx.doi.org/10.1016/j.soilbio.2009.07.008. Soil Biology and Biochemistry. 2009;41(11):2257–2271. doi: 10.1016/j.soilbio.2009.07.008.

Buckley Thomas R., James Sam, Allwood Julia, Bartlam Scott, Howitt Robyn, Prada Diana. Phylogenetic analysis of New Zealand earthworms (Oligochaeta: Megascolecidae) reveals ancient clades and cryptic taxonomic diversity. http://dx.doi.org/10.1016/j.ympev.2010.09.024. Molecular Phylogenetics and Evolution. 2011;58(1):85–96. doi: 10.1016/j.ympev.2010.09.024. [PubMed] [CrossRef] [Google Scholar]

Bunke D. Zur Morphologie und Systematik der Aeolosomatidae Beddard 1895 und Potamodrilidae nov. fam (Oligochaeta) Zool. Jahrb. Syst. 1967;94:187–368. [Google Scholar]

Bunke Dieter. Ultrastructure of the spermatozoon and spermiogenesis in the interstitial annelid *Potamodrilus fluviatilis*. http://dx.doi.org/10.1002/jmor.1051850206. Journal of Morphology. 1985;185(2):203–216. doi: 10.1002/jmor.1051850206. [PubMed] [CrossRef] [Google Scholar]

Bunke Dieter. Ultrastructural investigations on the spermatozoon and its genesis in *Aeolosoma litorale* with considerations on the phylogenetic implications for the aeolosomatidae (annelida) http://dx.doi.org/10.1016/0889-1605(86)90035-2. Journal of Ultrastructure and Molecular Structure Research. 1986;95:113–130. doi: 10.1016/0889-1605(86)90035-2. [CrossRef] [Google Scholar]

Carpenter D. .,, Sherlock E., Jones D. T., Chiminoides J., Writer T., Neilson R., Boag B., Keith A. M., Eggleton P. Mapping of earthworm distribution for the British Isles and Eire highlights the under-recording of an ecologically important group. Biodiversity and Conservation. 2012;21:475–485. doi: 10.1007/s10531-011-0194-x. [CrossRef] [Google Scholar]

Cech G, Dózsa-Farkas K. Identification of *Fridericia schmelzi* sp. n. combining morphological characters and PCR-RFLP analysis. In: Pop V. V., Pop A. A., editors. Advances in Earthworm Taxonomy II. Cluj University Press; Cluj-Napoca: 2005. 99–118 [Google Scholar]

Cech Gábor, Boros Gergely, Dózsa-Farkas Klára. Revision of *Bryodrilus glandulosus* (Dózsa-Farkas, 1990) and *Mesenchytraeus kuehnelti* Dózsa-Farkas, 1991 (Oligochaeta: Enchytraeidae) using morphological and molecular data. http://dx.doi.org/10.1016/j.jcz.2011.09.005. Zoologischer Anzeiger–A Journal of Comparative Zoology. 2012;251(3):253–262. doi: 10.1016/j.jcz.2011.09.005. [CrossRef] [Google Scholar]

Cech G., Csuzdi C., Marialigeti K. Remarks on the molecular phylogeny of the genus *Dendrobaena* (sensus Pop 1941) based on the investigation of 18S rDNA sequences. In: Pop V. V., Pop A. A., editors. Advances in Earthworm Taxonomy II (Annelida: Oligochaeta) Cluj University Press; Cluj-Napoca, Romania: 2005. 85–98 [Google Scholar]

Čekanovskaya O. V. Aquatic Oligochaeta of the U.S.S.R. Akademia Nauk S.S.S.R.; Moskva-Leningrad: 1962. 411. Russian. [Google Scholar]

Černosvitov L. System der Enchytraeiden. Bull. Ass. Russe Rech. Sci. Prague. 1937;5:263–295. [Google Scholar]

Chang Chih-Han, James Samuel. A critique of earthworm molecular phylogenetics. http://dx.doi.org/10.1016/j.pedobi.2011.07.015. Pedobiologia. 2011;54:S3–S9. doi: 10.1016/j.pedobi.2011.07.015.

Christensen Bent, Dózsa-Farkas Klára. Invasion of terrestrial enchytraeids into two postglacial tundras: North-eastern Greenland and the Arctic Archipelago of Canada (Enchytraeidae, Oligochaeta) http://dx.doi.org/10.1007/s00300-005-0076-3. Polar Biology. 2005;29(6):454–466. doi: 10.1007/s00300-005-0076-3. [CrossRef] [Google Scholar]

Christensen B., Dózsa-Farkas K. A new genus *Globulidrilus* and three new enchytraeid species (Oligochaeta: Enchytraeidae) from Seoraksan National Park (Korea) Journal of Natural History. 2012;46:2769–2785. doi: 10.1080/00222933.2012.737038. [CrossRef] [Google Scholar]

Christoffersen M. L. Species diversity and distributions of microdrile earthworms (Annelida, Clitellata, Enchytraeidae) from South America. Zootaxa. 2009;2065:51–68. [Google Scholar]

Coates K. A. Redescription of the oligochaete genus *Propappus*, and diagnosis of the new family Propappidae (Annelida: Oligochaeta) Proceedings of the Biological Society of Washington. 1986;99:417–428. [Google Scholar]

Cobolli Sbordoni M., De Matthaeis E., Alonzi A., Mattoccia M., Omodeo P., Rota E. Speciation, genetic divergence and palaeogeography in the Hormogastridae. Soil Biology and Biochemistry. 1992;24:1213–1221. doi: 10.1016/0038-0717(92)90096-g. [CrossRef] [Google Scholar]

Collado R., Hasprzak P, Schmelz R. M. Oligochaeta and Aphanoneura in two Northern German hardwater lakes of different trophic state. Hydrobiologia. 1999;406:143–148. doi: 10.1007/978-94-011-4207-6_14. [CrossRef] [Google Scholar]

Csuzdi Cs. Über ein Vorkommen von *Microscolex phosphoreus* (Dugès, 1837) (Oligochaeta: Acanthodrilidae) in Ungarn. Opusc. Zool. Budapest. 1986;22:63–66. [Google Scholar]

Csuzdi Cs. A catalogue of Benhamiinae species (Annelida: Oligochaeta: Acanthodrilidae) Ann. Naturhist. Mus. Wien. 1995;97:99–123.

Domínguez Jorge, Edwards Clive A., Dominguez John. The biology and population dynamics of *Eudrilus eugeniae* (Kinberg) (Oligochaeta) in cattle waste solids. http://dx.doi.org/10.1078/0031-4056-00091. Pedobiologia. 2001;45(4):341–353. doi: 10.1078/0031-4056-00091. [CrossRef] [Google Scholar]

Domínguez Jorge, Aira Manuel, Breinholt Jesse W., Stojanovic Mirjana, James Samuel W., Pérez-Losada Marcos. Underground evolution: New roots for the old tree of lumbricid earthworms. http://dx.doi.org/10.1016/j.ympev.2014.10.024. Molecular Phylogenetics and Evolution. 2015;83:7–19. doi: 10.1016/j.ympev.2014.10.024. [PMC free article] [PubMed] [CrossRef] [Google Scholar]

Dózsa-Farkas K, Felföldi T. Unexpected occurrence of *Hemifridericia bivesiculata* Christensen & Dózsa-Farkas, 2006 in Hungary, a species presumed to be endemic to Devon Island, Canada, and its comparative analysis with *Hemifridericia parva* Nielsen & Christensen, 1959 (Enchytraeidae, Oligochaeta) http://dx.doi.org/10.11646/zootaxa.3914.2.8. Zootaxa. 2015;3914(2):185–194. doi: 10.11646/zootaxa.3914.2.8. [PubMed] [CrossRef] [Google Scholar]

Dózsa-Farkas K., Hong Y. Three new *Hemienchytraeus* species (Enchytraeidae, Oligochaeta, Annelida) from Korea, with first records of other enchytraeids and terrestrial polychaetes (Annelida) Zootaxa. 2010;2406:29–56. [Google Scholar]

Dózsa-Farkas K., Schlaghamerský J. *Hrabeiella periglandulata* (Annelida: "Polychaeta") – Do apparent differences in chaetal ultrastructure indicate the existence of several species in Europe? Acta Zoologica Academiae Scientiarum Hungaricae. 2013;59:143–156. [Google Scholar]

Dózsa-Farkas K., Felföldi T., Hong Y. New enchytraeid species (Enchytraeidae, Oligochaeta) from Korea. Zootaxa. 2015;4006(1):2769–2785. doi: 10.11646/zootaxa.4006.1.9. [PubMed] [CrossRef] [Google Scholar]

Dózsa-Farkas K., Porco D., Boros G. Are *Bryodrilus parvus* Nurminen, 1970 and *Bryodrilus librus* (Nielsen and Christensen, 1959) (Annelida: Enchytraeidae) really different species? A revision based on DNA barcodes and morphological data. Zootaxa. 2012;3276:38–50.

Erséus Christer, Rota Emilia, Matamoros Lisa, Wit Pierre De. Molecular phylogeny of Enchytraeidae (Annelida, Clitellata) http://dx.doi.org/10.1016/j. ympev.2010.07.005. Molecular Phylogenetics and Evolution. 2010;57(2):849–858. doi: 10.1016/j.ympev.2010.07.005. [PubMed] [CrossRef] [Google Scholar]

Fauchald K., Rouse G. Polychaete systematics: Past and present. http://dx.doi. org/10.1111/j.1463-6409.1997.tb00411.x. Zoologica Scripta. 1997;26(2):71–138. doi: 10.1111/j.1463-6409.1997.tb00411.x. [CrossRef] [Google Scholar]

Fender W. M. Native earthworms of the Pacific Northwest: an ecological overview. In: Hendrix P. F., editor. Earthworm Ecology and Biogeography in North America. CRC Press Lewis Publ.; Florida: 1995. 53-66 [Google Scholar]

Fernández-Marchán D., Fernández R., Novo M., Díaz-Cosín D. New light into the hormogastrid riddle: morphological and molecular description of *Hormogaster joseantonioi* sp. n. (Annelida, Clitellata, Hormogastridae) http://dx.doi.org/10.3897/ zookeys.414.7665. ZooKeys. 2014;414:1–17. doi: 10.3897/zookeys.414.7665. [PMC free article] [PubMed] [CrossRef] [Google Scholar]

Gorgoń Szymon, Krodkiewska Mariola, Świątek Piotr. Ovary ultrastructure and oogenesis in *Propappus volki* Michaelsen, 1916 (Annelida: Clitellata) http://dx.doi. org/10.1016/j.jcz.2015.05.006. Zoologischer Anzeiger–A Journal of Comparative Zoology. 2015;257:110–118. doi: 10.1016/j.jcz.2015.05.006. [CrossRef] [Google Scholar]

Graefe U., Beylich A. First record of the aquatic earthworm *Sparganophilus tamesis* Benham, 1892 (Clitellata, Sparganophilidae) in Germany; Proceedings 5th Int. Oligochaete Taxonomy Meeting; Beatenberg, Switzerland. 2011. 25-25

CHAPTER IX:

SUB-PHYLUM—CRUSTACEA – "OUR COLORFUL COUSIN SEND SOME GREETINGS!"

"Once we see, however, that the probability of life originating at random is so utterly minuscule as to make it absurd, it becomes sensible to think that the favourable properties of physics on which life depends are in every respect deliberate….It is therefore almost inevitable that our own measure of intelligence must reflect …higher intelligences…even to the limit of God…such a theory is so obvious that one wonders why it is not widely accepted as being self-evident. The reasons are psychological rather than scientific." Sir Fred Hoyle

Animals that belong to the Subphylum, Crustacea, are part of a larger Phylum termed Arthropoda. The group of animals that belongs to the Subphylum, Crustacea, includes crabs, lobsters, shrimps, crayfish, prawns and barnacles. These animals have been in existence for 500 million years and appeared during the Cambrian Period. The Subphylum, Crustacea, contains about 67,000 species and most of the species are consumed by humans. The lobster was utilized for this study (see Figure 9.1).

Figure 9.1: Picture of Lobsters.

General Features

The lobster's body is comprised of segments that are grouped into three regions – the head or cephalon, the thorax or pereon, and the abdomen or pleon. Anterior to the thorax is the head with two pairs of antennae and three pairs of maxillipeds for feeding. There is also a pair of compound eyes carried by a pair of stalk, located in the head region. The thorax, which is covered by the carapace, and the head, are referred to as the cephalothorax region. The abdominal region carries six pairs of appendages. The first five pairs are called the pleopods, or, swimmerets. The last pair, called the uropod, has become modified to form the tail fan and the terminal telson. Most of these animals are bottom dwellers, and when they move, they could crawl, or, glide through the substratum. They do swim generally with the help of the pleopods.

Circulatory System, Gas Exchange and Excretion

Lobster has an open blood vascular system. The dorsal heart has the pericardial cavity surrounding it. Blood leaves the pericardial cavity and flows into the heart through the small lateral openings called ostia. The contraction of the heart usually closes all the ostia in the segments, and send blood out of the heart through the arteries to all tissue spaces in the body where gas exchanges for metabolic activities takes place. The blood is then drained from tissue spaces through a system of larger sinuses and returned to the pericardial cavity back to the heart.

The gills are the organs of gas exchange and air absorbed from the water in the gills are directly absorbed into the vascular system. The antennal glands or, green glands are the primary excretory organs. These organs, which are sac-like and well-supplied by blood, are located in the head region. There is an opening at the base of the second antenna, which opens to the outside for elimination of waste materials.

Nutrition

Lobsters are mostly predators and scavengers. The chelipeds are used for the collection, sorting, scraping tearing, cutting and crushing of food. The food once ingested is passed to the foregut which is a large chitinous cardiac stomach containing a dorsal and two lateral teeth, which forms the gastric mill. This leads posteriorly to the pyloric stomach, which ventrally contain opening of large extensive pair of digestive glands called the hepatopancreas, or midgut. The cardiac stomach, which acts as both the crop and the gizzard, uses the gastric mill to grind the food. Enzymes from the digestive glands are produced which helps to break the food into fine fluid materials. The digested food materials are then conveyed through channels to the digestive glands where absorption takes place into the vascular system. There are screens of setae along these channels that prevent solid and coarse particles from passing through. The undigested food matter is passed along to the intestine where it is expelled through the anus to the outside.

Reproduction

Most of the species in the Subphylum, Crustacea, have separate sexes. The male have a pair of testes, which are located in the dorsal part of the thorax. The sperm ducts open to the outside through the base of the fifth pair of legs. The females do have a pair of ovaries in the dorsal part of the thorax. The oviducts open to the outside through the gonophores at the base of the third pair of legs. When copulating, the male mounts the female and use the highly specialized anterior two pairs of the pleopods to transfer sperms into female gonophores. Fertilization could be internal, or, fertilized externally as the eggs are being passed out of the oviducts. The fertilized eggs are passed to the ventral surface of the abdomen and are attached to the pleopods by adhesive materials where the eggs are kept in a brood. The fertilized eggs develop into the larva stage and to a much advance larval stage called the zoea. After several molting or development stages called instars, the young adult lobsters develop. These instars leave from the ventral surface of the female adult abdomen to settle at the bottom of the water, and start new life cycles (see Figure 9.2).

Figure 9.2: Picture of Crabs.

Experiment with Mutagens

The study entails the exposure of the adult stages, larval stages and germ cell stages of the lobster to low intensity of radioactive elements, x-rays, and ultraviolet rays for periods of three months, six months, one year, and three years. The observations revealed that the ultraviolet rays did not affect the organism's development. Cultures exposed to x-rays and radioactive solutions were tolerated by the organisms. Beside occasional changes in color, there were no mutant varieties. Under medium intensity of the mutagens, the lobsters did survive for about three months. There were no observed mutant varieties. The germ cells, after fertilization, developed normally, but did not develop to the larval stage. Under high intensity of the mutagens, the organism did not survive past ten days. The germ cells that were exposed to the high dose of mutagens could not be fertilized, and there were no observed mutant varieties. Again, changes in the organism were limited to only color changes.

Genetic Mutations

Examinations were made of cells from each of the cultures with varying degrees of exposure to these mutagens. The findings revealed that there were no changes in the chromosomal content. The genome size of 1.2×10^9 bp remained constant. The gene size of approximately 11.1KB was consistent. The gene number of about 7.800 remained consistent. There were no visibly phenotypical changes in the organisms. It is possible that the changes in color observed could be the effect of different coding with regards to the production of the proteins affecting coloration. Also, it was observed that a new organelle associated with the chromosomes, termed the MasterCodon, was exerting control over phenotypical and genotypical changes.

Conclusion

Lobsters were affected by high level intensity of mutagens which resulted in quick death of the organism. The organisms' resistance to mutation is similar to other organisms in this study. The MasterCodon seems coded in such a unique manner to enable it to send messages to the genes for the control, maintenance, stabilization and preservation of cells' constancy and uniqueness in the organisms. It also sends coded messages to the genes to allow production of specific proteins for adaption to changes in the organism's environment. Such messages are also sent to prevent mutations at any cost, even if it means shutting-down the whole organism's system, resulting in death. The MasterCodon seems to be a life giver and preserver, and also the presenter of death.

Suggested Reading

Ahyong ST, Baba K, Macpherson E, Poore GCB. (2010) A new classification of the Galatheoidea (Crustacea: Decapoda: Anomura). Zootaxa 2676: 57–68. 10.11646/zootaxa.2676.1.4 [CrossRef] [Google Scholar]

Baba K. (1969) Four new genera with their representatives and six new species of the Galatheidae in the collection of the Zoological Laboratory, Kyushu University, with redefinition of the genus *Galathea*. Ohmu 2: 1–32. [Google Scholar]

Baba K. (1988) Chirostylid and galatheid crustaceans (Decapoda: Anomura) of the "Albatross" Philippine Expedition, 1907–1910. Researches on Crustacea, Special Number 2: 1–203. 10.18353/rcrustaceasn.2.0_1 [CrossRef] [Google Scholar]

Baba K, Macpherson E, Poore GCB, Ahyong ST, Bermudez A, Cabezas P, Lin C-W, Nizinski M, Rodrigues C, Schnabel K. (2008) Catalogue of squat lobsters of the world (Crustacea: Decapoda: Anomura – families Chirostylidae, Galatheidae and Kiwaidae). Zootaxa 1905: 1–220. 10.11646/zootaxa.1905.1.1 [CrossRef] [Google Scholar]

Baba K, Macpherson E, Lin CW, Chan T-Y. (2009) Crustacean Fauna of Taiwan: squat lobsters (Chirostylidae and Galatheidae). Taipei: National Science Council, Taiwan, R.O.C, ix + 312 pp.

Cabezas P, Macpherson E. (2014) A new species of *Paramunida* Baba, 1988 from the Central Pacific Ocean and a new genus to accommodate *P. granulata* (Henderson, 1885). ZooKeys 425: 15–32. 10.3897/zookeys.425.7882 [PMC free article] [PubMed] [CrossRef] [Google Scholar]

Cabezas P, Macpherson E, Machordom A. (2010) Taxonomic revision of the genus *Paramunida* Baba, 1988 (Crustacea: Decapoda: Galatheidae): a morphological and molecular approach. Zootaxa 2712: 1–60. 10.11646/zootaxa.2712.1.1 [CrossRef] [Google Scholar]

Cabezas P, Lin CW, Chan TY. (2011) Two new species of the deep-sea squat lobster genus *Munida* Leach, 1820 (Crustacea: Decapoda: Munididae) from Taiwan: morphological and molecular evidence. Zootaxa 3036: 26–38. 10.11646/zootaxa.3036.1.2 [CrossRef] [Google Scholar]

Cabezas P, Sanmartín I, Paulay G, Macpherson E, Machordom A. (2012) Deep under the sea: unraveling the evolutionary history of the deep-sea squat lobster *Paramunida* (Decapoda, Munididae). Evolution 66: 1878–1896. 10.1111/j.1558-5646.2011.01560.x

Folmer O, Black M, Hoeh W, Lutz R, Vrijenhoek R. (1994) DNA primers for amplification of mitochondrial cytochrome c oxidase subunit I from diverse metazoan invertebrates. Molecular Marine Biology and Biotechnology 3: 294–299. [PubMed] [Google Scholar]

Henderson JR. (1885) Diagnoses of new species of Galatheidae collected during the "Challenger" expedition. Annals and Magazine of Natural History (ser. 5) 16: 407–421. 10.1080/00222938509459908 [CrossRef]

Huelsenbeck JP, Ronquist F. (2001) MRBAYES: Bayesian inference of phylogenetic trees. Bioinformatics 17: 754–755. 10.1093/bioinformatics/17.8.754 [PubMed] [CrossRef] [Google Scholar]

Katoh K, Misawa K, Kuma KI, Miyata T. (2002) MAFFT: a novel method for rapid multiple sequence alignment based on fast Fourier transform. Nucleic Acids Research 30: 3059–3066. 10.1093/nar/gkf436 [PMC free article] [PubMed] [CrossRef] [Google Scholar]

Larsson A. (2014) AliView: a fast and lightweight alignment viewer and editor for large datasets. Bioinformatics 30: 3276–3278. 10.1093/bioinformatics/btu531 [PMC free article] [PubMed] [CrossRef] [Google Scholar]

Machordom A, Araujo R, Erpenbeck D, Ramos MA. (2003) Phylogeography and conservation genetics of endangered European Margaritiferidae (Bivalvia: Unionoidea). Biological Journal of the Linnean Society 78: 235–252. 10.1046/j.1095-8312.2003.00158.x [CrossRef] [Google Scholar]

Machordom A, Macpherson E. (2004) Rapid radiation and cryptic speciation in galatheid crabs of the genus *Munida* and related genera in the South West Pacific: molecular and morphological evidence. Molecular Phylogenetics and Evolution 33: 259–279. 10.1016/j.ympev.2004.06.001 [PubMed] [CrossRef] [Google Scholar]

Macpherson E. (1993) CrustaceaDecapoda: species of the genus *Paramunida* Baba, 1988 (Galatheidae) from the Philippines, Indonesia and New Caledonia. In:

Crosnier A. (Ed.) Résultats des Campagnes MUSORSTOM, volume 10.Mémoires du Muséum National d'Histoire Naturelle, Paris 156: 443–473.

Macpherson E, Baba K. (2011) Chapter 2. Taxonomy of squat lobsters. In: Poore GCB, Ahyong ST, Taylor J. (Eds) The biology of squat lobsters.CSIRO Publishing, Melbourne and CRC Press, Boca Raton, 39–71.

Macpherson E, Rodríguez-Flores PC, Machordom A. (in press) Squat lobsters of the families Munididae and Munidopsidae from Papua New Guinea. In: Ahyong ST, Chan T-Y, Corbari L (Eds) Tropical Deep-Sea Benthos 31, Papua New Guinea. Muséum national d'Histoire naturelle, Paris: 00-00 (Mémoires du Muséum national d'Histoire naturelle, 213).

McCallum AW, Cabezas P, Andreakis N. (2016) Deep-sea squat lobsters of the genus *Paramunida* Baba, 1988 (Crustacea: Decapoda: Munididae) from north-western Australia: new records and description of three new species. Zootaxa 4173: 201–224. 10.11646/zootaxa.4173.3.1 [PubMed] [CrossRef] [Google Scholar]

Miller MA, Pfeiffer W, Schwartz T. (2010) Creating the CIPRES Science Gateway for inference of large phylogenetic trees. Proceedings of the Gateway Computing Environments Workshop (GCE), 14 Nov. 2010, New Orleans, LA, 1–8. 10.1109/GCE.2010.5676129 [CrossRef]

Palumbi S, Martin A, Romano S, McMillan WO, Stice L, Grabowski G. (2002) The simple fool's guide to PCR. University of Hawaii, Honolulu, 45 pp. [Google Scholar]

Rambaut A. (2014) FigTree 1.4. 2 software. Institute of Evolutionary Biology, Univ. Edinburgh.

Rodríguez-Flores PC, Machordom A, Abelló P, Cuesta JA, Macpherson E. (2019) Species delimitation and multi-locus species tree solve an old taxonomic problem for European squat lobsters of the genus *Munida* Leach, 1820. Marine Biodiversity 49: 1751–1773. 10.1007/s12526-019-00941-3

Schnabel KE, Ahyong ST. (2019) The squat lobster genus *Phylladiorhynchus* Baba, 1969 in New Zealand and eastern Australia, with description of six new species. Zootaxa 4688: 301–347. 10.11646/zootaxa.4688.3.1 [PubMed] [CrossRef] [Google Scholar]

Schnabel KE, Cabezas P, McCallum A, Macpherson E, Ahyong ST, Baba K. (2011) Chapter 5. World-wide distribution patterns of squat lobsters. In: Poore GCB, Ahyong ST, Taylor J. (Eds) The biology of squat lobsters.CSIRO Publishing: Melbourne and CRC Press, Boca Raton, 149–182.

Swofford DL. (2002) PAUP*: phylogenetic analysis using parsimony, version 4.0 b10.

CHAPTER X:

PHYLUM, ARTHROPODA-SUBPHYLUM, UNIRAMIA "OUR WINGED COUSINS REVISITED"

"Unfortunately many scientists and non-scientists have made Evolution into a religion, something to be defended against infidels. In my experience, many students of biology – professors and textbook writers included – have been so carried away with the arguments for Evolution that they neglect to question it. They preach it ... College students, having gone through such a closed system of education, themselves become teachers, entering high schools to continue the process, using textbooks written by former classmates or professors. High standards of scholarship and teaching break down. Propaganda and the pursuit of power replace the pursuit knowledge. Education becomes a fraud." George Kocan

The sub-phylum, Uniramia, consist of arthropods that belong to animal groups like the centipedes, millipedes and insects. They contain the largest species diversification among the arthropods. They all possess a pair of antennae. They have their segmented appendages modified into legs. These groups of animals are believed to have evolved during the Mississippian period about 360 million years ago. For our study, we chose the class, Hexapoda, which includes insect with special emphasis on the common cockroach. The class, Hexapoda, is the largest group of animal with over 700,000 species inhabiting every type of terrestrial habitat with some even living in fresh waters. The development of wings for flying, which are great adaptation to terrestrial living, earned them the right to be called the first flying animals. This adaptation has given this group of animals the ability to escape dangers from predators. Also, to be able to gain access to favorable environment and food items. The development of a waxy cuticle, which acts as a water repellant and for the protection of water loss, allows the insects to live and colonize various territorial habitats (see Figure 10.1). For this study, cockroaches were utilized.

Figure 10.1: Figure of an Insect – a butterfly.

Body Structure and Movement

The Cockroach's body comprises of a head, a thoracic region and an abdominal region. On the lateral side of the head is a pair of antennae and a pair of compound eyes; and between them are small, simple eyes termed the ocelli. On the lateral – ventral side of the head are pair of mandibles and two pairs of maxillae. The second pair are fused together to form the longer lip, or, the labium. The upper lip or, labrum is formed from the projection of the head. At the base of the labrum is another median projection from the base of the oral cavity called the hypopharynx. The thoracic region is comprised of three segments with the last two segments carrying a pair of wings each. Each of the thoracic segments bears a pair of legs. The abdominal regions comprises of about 11 segments. The insect's wings are composed of hollow thickenings called veins, and these contain blood, nerves and trachea. Movement is generally by flying and only one pair of wings is used for this function. Also, insects do walk by the utilization of their legs which are meant for short distances.

Circulatory System, Gas Exchange and Excretion

The Cockroach's heart consists of a long abdominal tube with nine pairs of ostia. There is also an anterior aorta supplying blood to the head and thorax regions. The blood is pumped through tubes to tissues and is returned from these tissues through ventral vessels back of the heart. The gas exchange organs are the trachea with paired openings, or, percales, located in the last two thoracic segments, and the first eight abdominal segments. Gas intake go through the trachea, and the spiracles, directly into the vascular system, and then to the tissues in all segments, including the wings. The Malpighian tubules are the primary excretory organs. They are blind- ending tubules bathed with blood. These tubules open directly into the intestine where waste from the blood like uric acid, potassium, and sodium are deposited. Before passing out waste out of the rectum, more water is absorbed back at the rectum producing pasty and concentrated urine which is passed out with the feces through the anus.

Nutrition

The cockroach has become modified to feed on plant materials and other microphagous matter. The mouth parts, that include the cerium and labrum, have become adapted for sucking, cutting, piercing the biting. The food once injected in the mouth is digested by excretions from the salivary glands that help in the digestion of sugar and pectin, it also helps to lubricate the food and act as an anticoagulant. Then, the food moves to the pharynx, the crop and proventriculus (which comprise the foregut). The crop and proventriculus act as a grinding mill for the food. The mixed food materials move into the midgut, the stomach, or the ventriculus where enzymes and peritrophic secretions are secreted to help complete digestion. Gastric ceca, which are extensive sacs form the walls of the mid-gut, are used for absorption of the digested food. The undigested food wastes are sent to the hind gut, which comprises of the intestines and the rectum. The food wastes are passed out through the rectum to the anus before being expelled out as feces. Also, the rectum is the point of water absorption where the deposited urine is expelled with other excretory wastes.

Reproduction

The male reproductive organs which are located in the abdomen comprises of the paired testes, paired sperm ducts, paired seminal vesicles, paired accessory glands, and the ejaculatory duct which leads into the penis, or aedeagus. The female cockroach has a pair of ovaries located in the abdomen. These lead to a pair of lateral oviducts, and then to a common oviduct. The common oviduct opens to the vagina, which opens to the outside at the posterior end of the abdomen. Pair of accessory glands and the spermatheca opens into the vagina. Fertilization is internal. During copulation, the male use clasping structure at the posterior end of the abdomen to hold the female and insert the aedeagus in the vagina where the sperm are stored in the female spermatheca. As the eggs passed through the oviduct, they become fertilized by the sperm through an opening in the egg called micropule. The eggs are passed out from the vagina in batches and which are cemented together by secretions from the accessory glands. The female cockroach use terminal structures at the end of the abdomen called ovipositor, to deposit these batches of eggs in soil, plants, excrements and other decaying matter. The fertilized eggs undergo series of instars, or metamorphosis from larval nymphs' mode to adult cockroaches (see Figure 10.2).

Figure 10.2: Picture of bees.

Experiment with Mutagens

The common house cockroaches were exposed to varying cultures of low intensity radioactive elements, x-rays, and ultraviolet rays. The cultures contain adult stages, nymph stages, zygote stages and the germ cell. Also, the cultures were exposed to mutagens for periods of one-month to three years. Under low intensity of the mutagens, the cockroaches continued to thrive. Normal metabolic activity was not affected. Mutant varieties were not observed. Under medium intensity, the cockroaches, adapted very well. Within three months, the cultures exposed to both x-rays and radioactive elements became erratic. There were very aggressive and feeding habits became increasingly more rapid. Eventually they became comatose. Some mutant varieties were noticed form germ cells exposed to mutagens after fertilization. Some mutants were wingless. Some had missing appendages. Some had fewer segments with light brownish discolorations. All the mutant varieties did not survive past three months. The cultures exposed to higher intensity of mutagens did not survive pass one week. The fertilized germ cells did not develop into the larval stages.

Genetic Mutations

Observations on the organism's cells did not produce changes in the chromosomal content of the cockroaches. Even the mutant varieties did not show changes in the chromosomal content. The genome of 1.2×10^8 bp was unchanged. The genetic size was unchanged. The gene size of 11.2 KB was unchanged. The total number of genes of about 8,000 remained constant. There was an observation of a new organelle associated with the chromosomes that was termed the MasterCodon which prevents genotypical changes to the organism.

Conclusion

The study of the cockroach reveals that the organism resisted any attempt to mutate. The observed mutant varieties were adaptations to changes in the organism's new environment which were caused by the effect of the mutagens. The fact that the mutant varieties did not survive is adequate evidence that mutations were constantly rejected by the organism. The changes in body shapes, abnormally appendages and color were due to nucleotides bases shift with different proteins that are departures form the normal proteins. The new observed organelle, termed the MasterCodon, exerts significant influence and control on the genes so that there will be no genotypical mutations. Also, it ensures that the organism maintains its cellular constancy from generation to generation; and any attempt to change the organism's genotype caused the system to shut down which resulted in the death of the organism.

Suggested Reading

Arango, C.P. (2009) New species and new records of sea spiders (Arthropoda: Pycnogonida) from deep waters in Western Australia. Zootaxa, 1977 (1), 1–20. https://doi.org/10.11646/zootaxa.1977.1.1

Arango, C.P. & Krapp, F. (2007) A new species of Anoplodactylus (Arthropoda: Pycnogonida) from the Great Barrier Reef and discussion on the A. tenuicorpus–complex. Zootaxa, 1435 (1), 19–24. https://doi.org/10.11646/zootaxa.1435.1.2

Arango, C.P. & Maxmen, A. (2006) Proboscis ornamentation as a diagnostic character for the Anoplodactylus californicus–digitatus complex (Arthropoda: Pycnogonida) with an example from the Anoplodactylus eroticus female. Zootaxa, 1311 (1), 51–64. https://doi.org/10.11646/zootaxa.1311.1.3

Bamber, R.N. (1983) Some deep water Pycnogonids from the north-east Atlantic. Zoological Journal of the Linnean Society, 77, 65–74

Bamber, R.N. (1992) Some Pycnogonids from the South China Sea. Asian Marine Biology, 9, 193–203.

Bamber, R.N. (1997) Pycnogonids (Arthropoda: Pycnogonida) from the Cape d'Aguilar Marine Reserve, Hong Kong. In: Morton, B. (Eds.), The Marine Flora and Fauna of Hong Kong and Southern China IV. Hong Kong University Press, Hong Kong, pp. 143–157.

Bamber, R.N. (2004) Pycnogonids (Arthropoda: Pycnogonida) from Taiwan, with description of three new species. Zootaxa, 458 (1), 1–12. https://doi.org/10.11646/zootaxa.458.1.1

Bamber, R.N. (2008) A new species of Pycnogonum (Arthropoda: Pycnogonida: Pycnogonidae) from Hong Kong. Journal of Natural History, 42 (9–12), 815–819. https://doi.org/10.1080/00222930701850463

Bamber, R.N., El Nagar, A. & Arango, C.P. (Eds.) (2020) Pycnobase: World Pycnogonida Database. Available from: http:// www.marinespecies.org/pycnobase (accessed 23 March 2020)

Bamber, R.N. & Thurston, M.H. (1995) The deep-water pycnogonids (Arthropoda: Pycnogonida) of the northeastern Atlantic Ocean. Zoological journal of the Linnean

Society, 115, 117–162. https://doi.org/10.1111/j.1096-3642.1995.tb02325.x Böhm, R. (1879a) Über die Pycnogoniden des Königl. zoologischen Museums in Berlin, insbesondere über die von S. M. S. 'Gazelle' mitgebrachten Arten. Monatsberichte der Königlichen Preussische Akademie des Wissenschaften zu Berlin, 1879, 170–195. Böhm, R. (1879b) Über zwei neue von Herrn Dr. Hilgendorf in Japan gesammelte Pycnogoniden. Sitzungsberichte der Gesellschaft Naturforschender Freunde zu Berlin, 1879 (4), 53–60.

Calman, W.T. (1923) Pycnogonida of the Indian Museum. Records of the Indian Museum, 25, 265–299.

Calman, W.T. (1927) Report on the Pycnogonida. The Transactions of the Zoological Society of London, 22 (3), 403–410. https://doi.org/10.1111/j.1096-3642.1927.tb00389.x Caullery, M. (1896) Pycnogonides. Résultats scientifiques de la campagne du Caudan dans le golfe de Gascogne, août–sept. Annales de Universite de Lyon, 26, 361–364. Child, C.A. (1992) Shallow–water Pycnogonida of the Gulf of Mexico. Memoirs of the Hourglass Cruises, 9 (1), 1–86. Clark, W.C. (1963) Australian Pycnogonida. Records of the Australian Museum, 26, 1–82. https://doi.org/10.3853/j.0067-1975.26.1963.669

Cole, L.J. (1909) Reports on the scientific results of the expedition to the Eastern Tropical Pacific in charge of Alexander Agassiz. by the US. Fish Commission Steamer Albatross from October 1904 to March 1905. XIX. Pycnogonida. Bulletin of the Museum of Comparative Zoology at Harvard College, 52 (11), 185–192.

Dietz, L., Dömel, J.S., Leese, F. Lehmann, T. & Melze, R.R. (2018) Feeding ecology in sea spiders (Arthropoda: Pycnogonida): what do we know? Frontiers in Zoology, 15, 1–16. https://doi.org/10.1186/s12983-018-0250-4

Dohrn, A. (1881) Die Pantopoden des Golfes von Neapel und der angrenzenden Meeres —Abschnitte. Monographie der Fauna und Flora des golfes von Neapel, 3, 1–252. https://doi.org/10.5962/bhl.title.16340 Fage, L. (1956) Sur deux espèces de pycnogonides du Sierra Leone. Bulletin du Muséum d'Histoire naturelle de Paris, Série 2, 28 (3), 290–294.

Hedgpeth, J.W. (1948) The Pycnogonida of the western North Atlantic and the Caribbean. Proceedings of the United States National Museum, 97, 157–342. https://doi.org/10.5479/si.00963801.97-3216.157

Hedgpeth, J.W. (1949) Report on the Pycnogonida collected by the Albatross in Japanese waters in 1900 and 1906. Proceedings of the United States National Museum, 98, 233–321. https://doi.org/10.5479/si.00963801.98-3231.233

Hilton, W.A. (1939) A preliminary list of pycnognids (sic) from the shores of California. Journal of Entomology and Zoology and Pomona College, 31 (2), 27–35.

Hodge, G. (1864) XIII.—List of the British Pycnogonoidea, with descriptions of several new species. The Annals and Magazine of Natural History, 13 (74), 113–117. https://doi.org/10.1080/00222936408681584 Hoek, P.P.C. (1881) Report on the Pycnogonida dredged by HMS Challenger 1873–76. Reports of the Scientific Results of the Exploring Voyage of HMS Challenger, 3 (10), 1–167.

Hong, J.S. & Kim, I.H. (1987) Korean pycnogonids chiefly based on the collections of the Korea Ocean Research and Development Institute. The Korean Journal of Systematic Zoology, 3, 137–164.

Huang, Z.G. (2008) Marine species and their distribution in China. China Ocean Press, Beijing, 1191 pp. [in Chinese]

Huang, Z.G. & Lin, M. (2012a) The living species in China's seas. Vol.2. China Ocean Press, Beijing, 748 pp. [in Chinese] Huang, Z.G. & Lin, M. (2012b) An illustrated guide to species in China's seas. Vol. 5. China Ocean Press, Beijing, 399 pp. [in Chinese]

Huang, Z.G. & Cai R.X. (1981) Marine biofouling and its prevention. Vol. 1. China Ocean Press, Beijing, 352 pp. [in Chinese] Ives, J.E. (1891) Echinoderms and arthropods from Japan. Proceedings of the Academy of Natural Sciences of Philadelphia, 43, 210–223. https://doi.org/10.1038/043223b0

Krapp, F., Kocak, C. & Katagan, T. (2008) Pycnogonida (Arthropoda) from the eastern Mediterranean Sea with description of a new species of Anoplodactylus. Zootaxa, 1686 (1), 57–68. https://doi.org/10.11646/zootaxa.1686.1.5

Lee, D. & Kim, W. (2020) New species of Pycnogonum (Pycnogonida: Pycnogonidae) from Green Island, Taiwan, with an additional note on the holotype of P. spatium. Zootaxa, 4750 (1), 122–130. https://doi.org/10.11646/zootaxa.4750.1.6

Li, R.G. (2003) Macrobenthos on the continental shelves and adjacent waters, China Sea. China Ocean Press, Beijing, 420 pp. [in Chinese] Liu, R.Y. (2008) Checklist of marine biota of China seas. Vol. 2. China Science Press, Beijing, 661 pp. [in Chinese]

Loman, J.C.C. (1911) Japanische Podosomata: Beiträge zur Naturgeschichte Ostasiens, herausgegeben von F. Doflein. Denkschriften der Kaiserlichen Akademie der Wissenschaften (Mathematisch-Naturwissenschaftliche Classe), Supplement 2 (4), 1–18.

Lou, T.H. (1936a) Sur deux nouvelles varietes de Pycnogonides recueillies a Tsing–Tao, dans la Baie de Kiao-Chow, Chine. Contributions from the Institute of Zoology, National Academy of Peiping, 3 (1), 1–34.

Lou, T.H. (1936b) Note sur Lecythorhynchus hilgendorfi Bohm (Pycnogonida). Contributions from the Institute of Zoology, National Academy of Peiping, 3 (5), 133–163.

Lucena, R.A., Araújo, J.P. & Christoffersen, M.L. (2015) A new species of Anoplodactylus (Pycnogonida: Phoxichilidiidae) from Brazil, with a case of gynandromorphism in Anoplodactylus eroticus Stock, 1968. Zootaxa, 4000 (4), 428–444. https://doi.org/10.11646/zootaxa.4000.4.2

Lucena, R.A. & Christoffersen, M.L. (2018a) An annotated checklist of Brazilian sea spiders (Arthropoda: Pycnogonida). Zootaxa, 4370 (2), 101–122. https://doi.org/10.11646/zootaxa.4370.2.1

Lucena, R.A. & Christoffersen, M.L. (2018b) Anoplodactylus (Pycnogonida: Phoxichilidiidae) from Brazil, new records and two new species. Turkish Journal of Zoology, 42 (4), 372–388. https://doi.org/10.3906/zoo-1712-1

Maxmen, A., Browne, W.E., Martindale, M.Q. & Giribet, G. (2005) Neuroanatomy of sea spiders implies an appendicular origin of the protocerebral segment. Nature, 437 (7062), 1144–1148. https://doi.org/10.1038/nature03984

Müller, H.G. & Krapp, F. (2009) The pycnogonid fauna (Pycnogonida, Arthropoda) of the Tayrona National Park and adjoining areas on the Caribbean coast of Colombia. Zootaxa, 2319 (1), 1–138. https://doi.org/10.11646/zootaxa.2319.1.1

Nakamura, K. & Child, C.A. (1983) Shallow-water Pycnogonida from the Izu Peninsula, Japan. Smithsonian Contributions to Zoology, 386, 1–71. https://doi.org/10.5479/si.00810282.386

Nakamura, K. & Child, C.A. (1991) Pycnogonida from waters adjacent to Japan. Smithsonian Contributions to Zoology, 512, 1–74. https://doi.org/10.5479/si.00810282.512

Ohshima, H. (1933) Pycnogonids taken with a tow-net. Annotationes Zoologicae Japonenses, 14 (2), 211–220. Ortmann, A.E. (1890) Bericht über die von Herrn Dr, Bericht Döderlein in Japan gesammelten Pycnogoniden. Zoologische Jahrbücher, Systematik, 5 (1), 157–168.

Schimkewitsch, W. (1893) Compte rendu sur les Pantopodes recueillis pendant les Explorations de I'Albatross en 1891. Bulletin of the Museum of Comparatiw Zoology, Haward, 25 (2), 27–43.

Schimkewitsch, W. (1913) Einige neue Pantopoden. Annuaire du Musee Zoologique de I'Acaddmie Impgrialedes Sciences de St-Petersbourg, 18, 240–248.

Shao, K.T., Peng, C.I. & Wu, W.J. (2010) Taiwan species checklist 2010. Forestry Bureau, Council of Agriculture, Executive Yuan, Taipei, 911 pp. [in Chinese]

Slater, H.H. (1879) On a new genus of Pycnogon and a variety of Pycnogonum littorale from Japan. The Annals and Magazine of Natural History, 3, 281–283. https://doi.org/10.1080/00222937908562400

Staples, D.A. (1979) Three new species of Propallene (Pycnogonida: Callipallenidae) from Australian waters. Transactions of the Royal Society of South Australia, 103, 85–93.

Stock, J.H. (1954) Papers from Dr. Th. Mortensen's Pacific Expedition 1914–1916. LXXVII. Pycnogonida from Indo-West-Pacific, Australian, and New-Zealand Waters. Videnskabelige Meddelelser fra Dansk Naturhistorisk Forening i Kjøbenhavn, 116, 1–168.

Stock, J.H. (1956) Pantopoden aus dem Zoologischen Museum Hamburg. Mitteilungen aus dem Hamburgischen Zoologischen Museum und Institut, 54, 33–48.

Stock, J.H. (1968) Pycnogonida collected by the Galathea and Anton Bruun in the Indian and Pacific Oceans. Videnskabelige Meddelelser Dansk Naturhistorisk Forening, 131, 7–65.

Stock, J.H. (1974) Pycnogonida from the Continental Shelf, Slope, and Deep Sea of the Tropical Atlantic and East Pacific. Bulletin of Marine Science, 24 (4), 957–1092.

Stock, J.H. (1975) The pycnogonid genus Propallene Schimkewitsch, 1909. Bulletin Zoologisch Museum, 4, 89–94. Stock, J.H. (1978) Abyssal Pycnogonida from the north-eastern Atlantic bassin part 1. Cahiers de biologie marine, 19, 189– 219.

Stock, J.H. (1979) Pycnogonida from the mediolittoral and infralittoral zones in the tropical western Atlantic. Studies on the Fauna of Curaçao and Other Caribbean Islands, 59, 1–32.

Stock, J.H. (1986) Pycnogonida from the Caribbean and the Straits of Florida. Bulletin of marine Science, 38, 399–441. Sun, S.Y. (2009) Species composition and distribution of pycnogonids in the Port of Kaohsiung. Marine Biotechnology & Resources, National Sun Yat-sen University, Kaohsiung, 69 pp. [in Chinese]

Sun, S.Y. & Chen, I.M. (2008) Taxonomic basic study on pycnogonida in port of Kaohsiung. 2008 Workshop: Research and Status of Taiwan Species Diversity, 2008, 220–223. [in Chinese]

Wang, J.J., Xia Z., Lin, R.C., Liang, Q.Y., Lin, H.S., Wang, J.J. & Zheng, C.X. (2015) A new species of Hemichela Stock, 1954 from the South China Sea (Arthropoda, Pycnogonida, Ammotheidae). ZooKeys, 526, 1–8. https://doi.org/10.3897/zookeys.526.5963

Williams, G. (1941) A Revision of the Genus Anoplodactylus with a New Species from Queensland. Memoirs of the Queensland Museum, 12, 33–39.

Wilson, E. (1881) Report on the Pycnogonida. Bulletin of the Museum of Comparative Zoology, Harvard, 8 (12), 239–256.

CHAPTER XI:

PHYLUM—ECHINODERMATA "FRIENDLY NEPHEWS ARE HERE"

"We are told dogmatically that Evolution is an established fact; but we are never told who has established it, and by what means. We are told, often enough, that the doctrine is founded upon evidence, and that indeed this evidence 'is henceforward above all verification, as well as being immune from any subsequent contradiction by experience;' but we are left entirely in the dark on the crucial question wherein, precisely, this evidence consists."
Smith, Wolfgang

The animals belonging to the phylum Echinodermata were thought to have evolved about 650 million years ago during the Ordovician period. These animals are usually radially symmetrical, but, the larvae are bilaterally symmetrical. The external part of the body is usually covered by spines and their endoskeletons contain ossicles that are calcified. There are over 7,000 existing species and 13,000 extinct species in this phylum; and virtually all of them are marine inhabitants. They include such animals as the sea lilies, sea cucumbers, sea urchins, sea stars, brittle stars and sand dunes. For this study, the sea star belonging to the sub-class asteroidean was chosen (see Figure 11.1).

Figure 11.1: Picture of a Sea Star.

Body Structure and Movement

The sea star has an outer layer of monociliated epithelium. Underneath this layer is the dermis, which secretes the endoskeleton that contains calcified ossicle. The endoskeleton is perforated by irregular canals. The main body has a central disc from which radiates about five arms. These animals generally exhibit a variety of colors. Below the dermis are muscles layers, which control the bending and stretching of the arms. The coelom, which is lined by ciliated peritoneum, carries coelomic fluid through the body and this forms the main internal transport system. The body is covered by spines, which are offshoots from the ossicle in the endoskeleton. Between these spines are pores for gas exchange and excretion. The mouth could be found in the oral surface of the central disc and generally faces downward. From the mouth radiates the ambulacral groove to all the arms which carry the tube feet, or podia. The anus is located on the smooth or vanular arboreal surface. Movement is by swimming and gliding by the use of muscle layer of the podium, and arms which contract for bending and stretching.

Circulatory System, Gas Exchange and Excretion

The vascular system is composed of pocket of body cell called the podia. This system opens to the outside through a button-like structure, called madreporite containing perforations of tiny canals. These tiny canals are about five pairs of projections called Tiedemann's bodies that function as a pathway between the water vascular system and the coelom. Radial canals from the ring canal extend to each of the lateral canals found at regular intervals, and which terminates in the ampulla, located in the body coelom and the podium. They also project into the ambulacral groove. The water vascular system helps in the function of locomotion, gas exchange, excretion and feeding. On the surface of the body are papulae, which allow oxygen, carbon dioxide and ammonia to be taken in and out from the surrounding water and circulated to the inner coelomic fluid. Wastes in the form of coelomocytes are passed through the vascular system to the papulae where they are excreted into the sea water.

Nutrition

The sea stars are generally carnivorous and scavengers. They feed mostly on mollusks, crustaceans and other echinoderms. The mouth is used for swallowing the prey or the cardiac stomach could be exerted to suck in the prey. The mouth at the oral side open into a large thick walled cardiac stomach, which covers internally most of the central disc for digestion of food. Digestion starts in the cardiac stomach from secretions from cells in its walls. The food matter then move into the aboral pyloric stomach. A pair of digestive glands located in each arm releases enzymes into both the cardiac stomach and the pyloric stomach and helps complete the digestion of the food products. The digested food is absorbed from the digestive glands directly into the coelomic fluid, or through radially arranged system of sinus channels called the hemal system. The undigested food matter or feces are passed to the intestine which has pockets called rectal ceca. The rectal ceca function as pump to expel the feces from the intestine through the anus located at the aboral surface of the animal.

Nervous System

The nervous system in the sea star is still very primitive. The epidermal cells are still poorly ganglionated. There is a nerve ring that encircles the mouth which contains radial nerves that extends to each of the arms. Messages are relayed through fibers in the nerve ring in conjunction with those in the radial nerves by conducting rapid information to neurons in the epidermal nerve plexus (see Figure 11.2).

Figure 11.2: Picture of Sea Stars.

Reproduction

The sexes in the sea stars are separate. Each of the male and female species has a pair of gonads in each arm. The gonophores in each arm open to the outside at the base of the arm. Fertilization is external. The eggs are shed freely and regularly into the surrounding water and they are fertilized immediately by the sperms. The fertilized eggs may develop within a brood, or in planktons, and the first larval stage emerges. The larval stage called bipinnaria is bilaterally symmetrical and develops larval arms, suckers and external ciliated bands as it grows. This larval stage is called the brachiolaria and normally settles down to the substratum. A complex metamorphosis stages occur with the little sea star crawling out and away from the larval remnants.

Experiment with Mutagens

Several cultures containing sea stars in adult forms, larval stages, and germ stages were exposed to various intensities of mutagen for periods of six months, one year, and three years. Under exposure to low intensity ultraviolet rays, x-rays, and radioactive solutions, the sea stars had good tolerance levels to the mutagens. The animal continued to thrive and normal metabolic activity was not affected. However, there were no mutant varieties. Similar cultures were simultaneously exposed to medium intensity of the mutagens. The sea star adapted rapidly and survived longer. During this stage, there was rapid increase in metabolic activities. The water vascular system became more active. There were more color changes in bright colors of red and burgundy. Feeding activities became more rapid. Eventually, the animal ceased all activities due to exhaustion and became comatose. Similar observations were recorded for the larval stages and they did survive long to develop into stunted young adults. The germ cells that were left in cultures exposed to the mutagens for one year produced a lot of mutant varieties. The mutant varieties came in all sorts or hue of different colors. Some had three and four fully developed arms with the other arms slightly deformed. Some had smaller bodies and the spines were poorly developed. It should be noted that these results were more prominent in cultures that were exposed to all three mutagens simultaneously. Under exposure of high intensity mutagens, the sea star did not survive past the tenth day. There were no observed mutant variations. The germ cell cultures, after attempts at fertilization, did not develop into the next stage.

Genetic Mutations

Observation of the effect of mutagens on the cellular nuclei revealed that there were no changes or variations in the chromosomal content of the organism. The mutant varieties did not produce changes in chromosomal number and the genome size remained constant. In all cells observed, the genome, the size, and the gene number remained virtually constant. The changes in colors; the deformities in the animal; and the size differentiation as observed in the mutant varieties were all due to shift in nucleotide bases. These bases are responsible for the genetic codes that provide information for the production of protein and enzymes. The point mutation occurring only changed the genetic codes thereby producing different proteins, hence, the observed changes. From the observed cells, a new organelle associated with the chromosomes in the nucleus termed the Master Codon was responsible for maintenance of cells' constancy.

Conclusion

The sea star did demonstrate a lot of adaptability to the mutagens under low intensity. It was observed that the organism did resist any attempt to mutate. All the mutant varieties did not survive. The MasterCodon is the organelle in the nucleus responsible for ensuring that no mutation occurs. It guarantees that the organisms' uniqueness and constancy is maintained from generation to generation. Any mutations occurring accidentally due to the effect of mutagens resulted in the death of the organism or the shutting down of all cells activities by the MasterCodon. The MasterCodon however, seems to favors and allow mutations that are favorable to the organism in it adaptation to it environment without changing or seriously affecting the uniqueness and constancy of the cell and ultimately the organism. Essentially, it prevents genotypical modifications.

Suggested Reading

Baker, A.N. (1979) Some Ophiuroidea from the Tasman Sea and adjacent waters. *New Zealand Journal of Zoology,* 6, 21–51.

Balinsky, J.B. (1957) The Ophiuroidea of Inhaca Island. *Annals of the Natal Museum* 14, 32.

Bell, F.J. (1884) Echinodermata. *In: Report on the zoological collections made in the Indo-Pacific Ocean during the voyage of H.M.S. "Alert", 1881-2.* Order of the Trustees, London, pp. 117–177, 509–512.

Bell, F.J. (1888) Descriptions of four new species of Ophiurids. *Proceedings of the Zoological Society of London.* 188, 281–284.

https://doi.org/10.1111/j.1469-7998.1888.tb06708.x

Bell, F.J. (1909) Report on the echinoderms (other than holothrians) collected by Mr. J. S. Gardiner in the western parts of the Indian Ocean. *Transactions of the Linnean Society of London 2nd Series Zoology,* 13, 17–22. https://doi.org/10.1111/j.1096-3642.1909.tb00406.x

Bergmann, W. (1900) Echinoderma für 1894. *Archiv für Naturgeschichte,* 35, 351–380.

Blainville, H.M. (1834) 1 FG Levrault *Manuel d'actinologie ou de zoophytologie.*

Bribiesca-Contreras, G., Solís-Marín, F.A., Laguarda-Figueras, A. & Zaldívar-Riverón, A. (2013) Identification of echinoderms (Echinodermata) from an anchialine cave in Cozumel Island, Mexico, using DNA barcodes. *Molecular Ecology Resources,* 13, 1137–1145. https://doi.org/10.1111/1755-0998.1209

Brock, J. (1888) Die Ophiuriden fauna des indischen Archipels. *Zeitschrift für wissenschaftliche Zoologie,* 47, 465–539.

Chang, F.Y., Liao, Y.L., Wu, B.L. & Chen, L. (1964) *Illustrated Fauna of China Echinodermata.* Science Press, Beijing.

Chao, S.-H., Chen, C.-P. & Chang, K.-H. (1991) Some shallow-water ophiurans (Echinodermata: Ophiuroidea) of Taiwan. *Bulletin of the Institute of Zoology, Academica Sinica,* 30, 117–126.

Cherbonnier, G. & Guille, A. (1978) 48 Faune de Madagascar *Echinodermes: Ophiurides*. Editions du Centre National de la Recherche Scientifique, Paris.

Clark, A.H. (1949) Ophiuroidea of the Hawaiian Islands. *Bernice P. Bishop Museum Bulletin*, 195, 1–133.

Clark, A.M. (1952) Some echinoderms from south Africa. *Transacations of the Royal Society of South Africa*, 33, 193.

https://doi.org/10.1080/00359195109519884

Clark, A.M. (1953) A revision of the genus *Ophionereis* (Echinodermata, Ophiuroidea). *Proceedings of the Zoological Society of London*, 123, 65–94.

Clark, A.M. (1965) Japanese and other ophiuroids from the collections of the Münich Museum. *Bulletin of the British Museum (Natural History) Zoology*, 13, 39–71.

Clark, A.M. (1967) Echinoderms from the Red Sea. Part 2. (Crinoids, Ophiuroids, Echinoids, and more Asteroids). *Sea Fisheries Research Station Israel Bulletin*, 41, 26–58.

Clark, A.M. (1970) Notes on the family Amphiuridae (Ophiuroidea). *Bulletin of the British Museum (Natural History) Zoology*, 19, 1–81.

https://doi.org/10.5962/bhl.part.24085

Clark, A.M. (1974) Notes on some echinoderms from southern Africa. *Bulletin of the British Museum (Natural History), Zoology*, 26, 421–487.

https://doi.org/10.5962/bhl.part.209

Clark, A.M. (1976) Asterozoa from Amsterdam and St Paul Islands, southern Indian Ocean. *Bulletin of the British Museum (Natural History)*, 30, 247–261.

Clark, A.M. (1980) Some Ophiuroidea from the Seychelles Islands and Inhaca, Mozambique. *Revue Zoologique Africaine (Bruxelles)*, 94, 534–558.

Clark, A.M. (1982) Echinoderms of Hong Kong. *In:* Morton, B. & Tseng, C.K. (Eds.), *Marine Fauna and Flora of Hong Kong*. Hong Kong University Press, Hong Kong, pp. 485–501.

Clark, A.M. & Courtman-Stock, J. (1976) *The echinoderms of Southern Africa.* British Museum (Naturak History), London.

Clark, A.M. & Melville, R.V. (1976) *Ophiolepis* Müller & Troschel, 1840, request for designation of a type-species under the plenary powers. Z.N.(S.) 2097. *The Bulletin of Zoological Nomenclature,* 32, 268–269.

Clark, A.M. & Rowe, F.W.E. (1971) 690 Trustees of the British Museum (Natural History) *Monograph of shallow-water Indo-West Pacific Echinoderms.* British Museum (Naturak History), London.

Clark, H.L. (1908) Some Japanese and East Indian echinoderms. *Bulletin of The Museum of Comparative Zoology,* 51, 279–311.

Clark, H.L. (1909) Notes on some Australian and Indo-Pacific Echinoderms. *Bulletin of the Museum of Comparative Zoölogy at Harvard College,* 52, 109–135.

Clark, H.L. (1911) North Pacific Ophiurans in the collection of the United States National Museum. *Smithsonian Institution United States National Museum Bulletin,* 75, 1–302.
https://doi.org/10.5479/si.03629236.75.1

Clark, H.L. (1915) Catalogue of recent ophiurans: based on the colletion of the Museum of Comparative Zoology. *Memorians of the Museum of Comparative Zoölogy at Havard College,* 25, 163–376.

Guille, A. & Vadon, C. (1986) Ophiurudae l'océan Indien profond. *Indo-Malayan Zoology,* 3, 167–188.

Iliffe, T.M. & Kornicker, L.S. (2009) Worldwide diving discoveries of living fossil animals from the depths of anchialine and marine caves. *Smithsonian Contributions to the Marine Sciences,* 38, 269–280.

Irimura, S. (1969) Supplementary report of Dr. Murakami's paper on the ophiurans of Amakusa, Kyushu. *Publications from the Amakusa Marine Biological Laboratory, Kyushu University,* 2, 37–48.

Irimura, S. (1979) Ophiuroidea of Sado Island, the Sea of Japan. *Annual Report of the Sado Marine Biological Station, Niigata University,* 9, 1–6.

Irimura, S. (1981) Ophiurans from Tanabe Bay and its vicinity, with the description of a new species of *Ophiocentrus*. *Publications of the Seto Marine Biological Laboratory,* 26, 15–49. https://doi.org/10.5134/176023

Irimura, S. (1982) *The brittle-stars of Sagami Bay, collected by His Majesty tje Emperor of Japan.* Biological Laboratory, Imperial Household, Japan.

Irimura, S. (1991) Ophiuroidea. *In:* F. R. C. Association (Ed.), *Echinoderms from Continental Shelf and Slope around Japan, 2.* Tosho Printing Co., Ltd, Tokyo, pp. 111–152.

Irimura, S. (1993) *Ophiostriatus sexradiatus*, a new species of Ophiuroidea from the North Mariana Islands. *Bulletin National Science Museum, Tokyo,* Series A 19, 161–164.

Irimura, S. & Tachikawa, H. (2002) Ophiuroids from the Ogasawara Islands (Echinodermata: Ophiuroidea). *Ogasawara Research,* 28, 1–27.

Ives, J.E. (1891) Echinoderms and Arthropods from Japan. *Proceedings of the Academy of Natural Sciences, Philadelphia,* 43, 210–223.

James, D.B. (1981) Studies on Indian echinoderms–8 on a new genus *Ophioelegans* (Oph1uroidea : Ophiuridae) With Notes on Ophiolepis superba H. L. Clark, 1938. *Journal of the Marine Biological Association of India,* 23, 15–18.

James, D.B. (1989) Echinoderms of Lakshadweep and their zoogeography. *CMFRI bulletin 43: Marine living resources of the union territory of Lakshadweep- an indicative survey with suggestions for development,* 43, 97–143.

Jeng, M.S. (1998) Shallow-water echinoderms of Taiping Island in the South China Sea. *Zoological Studies,* 37, 137–153.

Kakui, K. & Fujita, Y. (2018) *Haimormus shimojiensis* a new genus and species of Pseudozeuxidae (Crustacea: Tanaidacea) from a submarine limestone cave in Northwestern Pacific. *PeerJ,* 6, e4720. https://doi.org/10.7717/peerj.4720

Kikuchi, T. (1977) Biological survey of benthic macrofauna in Chijiwa Bay, west Kyushu II. Ophiuroidea. *Publications from the Amakusa Marine Biological Laboratory, Kyushu University,* 4, 127–141.

Kingston, S.C. (1981) The Swain Reefs Expedition: Ophiuroidea. *Records of the Australian Museum,* 33, 123–147. https://doi.org/10.3853/j.0067-1975.33.1980.277

Koehler, R. (1898) Echinoderms recueillis par l'Investigator dans l'Ocean Indien, II. Les Opiures littorales. *Bulletin scientifique de la France et de la Belgique,* 31, 55–126.

Koehler, R. (1900) Ophiures recueilles par "Investigator" dans l'Océan Indien. In: *Echinodermata of the Indian Museum. Ophiuroidea. Illustration of the shallow-water Ophiuroidea collected by the Royal Indian Marine Survey Ship Investigator.* Indian Museum, Calcatta, pp. 4, 22 pls.

Koehler, R. (1904) Ophiures de l'Expédition du Siboga. Part I. Ophiures de mer profonde. *Siboga-Expeditie,* 45a, 1–238.

Koehler, R. (1905) Ophiures de l'Expédition du Siboga. Part II. Ophiures de mer profonde. *Siboga-Expeditie,* 45b, 1–142.

Koehler, R. (1907a) Revision de la collection des Ophiures du Museum d'histoire Naturelle Paris. *Bulletin Scientifique de la Franca et de la Belgique,* 41, 279–351.

Koehler, R. (1907b) Ophiuroidea. *Fauna Südwest-Australiens,* 1, 241–254.

Koehler, R. (1922) Echinodermata: Ophiuroidea. Australasian Antarctic Expedition 1911–1914. *Scientific Report,* Series C, 8, 5–98.

Koehler, R. (1930) Ophiures recueillies par le Docteur Th. Mortensen dans les Mers d'Australie et dans l'Archipel Malais. Papers from Dr. Th. Mortensen's Pacific Expedition 1914–16. LIV. *Videnskabelige Meddelelser fra Dansk naturhistorisk Forening,* 89, 1–295.

Komai, T. & Fujita, Y. (2018) A new genus and new species of alpheid shrimp from a marine cave in the Ryukyu Islands, Japan, with additional record of *Salmoneus antricola* Komai, Yamada & Yunokawa, 2015. *Zootaxa,* 4269 (4), 575–586. https://doi.org/10.11646/zootaxa.4369.4.7

Lamarck, J.P.B.A., 1816. *Histoire naturelle des Animaux sans veretèbres, vol. 2, first ed.* Paris, pp. 522–568.

Liao, Y.L. (1978) The echinoderms of Xisha Islands Guangdong Province, China.2. Ophiuroidea. *Studia Marina Sinica,* 12, 69–104.

Liao, Y. (2004) Echinodermata Ophiuroidea. In: *Fauna Sinica*. Science Press, Beijing, pp. 505.

Liao, Y. & Clark, A.M. (1995) *The echinoderms of southern China.* Science Press, Beijing, New York.

Ljungman, A.V. (1867) Ophiuroidea viventia huc usque cognita enumerat. *In: Öfversigt af Kgl. Vetenskaps-Akademiens Förhandlingar.* pp. 221–272.

de Loriol, P. (1893a) Catalogue raisonné des Échinodermes recueillis par M. V. de Robillard à l'Ile Maurice. III. Ophiurides et Astrophytides. *Mémoires de la Société de Physique et D'Histoire Naturelle de Genève,* 32, 1–63.

de Loriol, P. (1893b) Echinodermes de la Baie d'Amboine. *Revue Suisse De Zoologie,* 1, 359–426.

Lütken, C.F. (1859) Additamenta ad historiam Ophiuridarum. Anden Afdelning. *Det kongelige danske Videnskabernes Selskabs Skrifter. 5 Raekke, Naturvidenskabelig og mathematisk Afdelning,* 5, 177–271.

Lütken, C.F. (1869) Additamenta ad historiam Ophiuridarum. Beskrivende og kritiske Bidrag til Kundskab om Slangestjernerne. Tredie Afdelning. *Videnskabernes Selskabs Skrifter. 5 Raekke, Naturvidenskabelig og mathematisk Afdelning,* 8, 20–109.

Lütken, C.F. (1872) Ophiuridarum novarum vel minus cognitarum descriptiones nonnullae. ogle nye eller mindre bekjerdte Slangestjernerne beskrevne – Med nogle Bemaerkninger om Selvdelingen hos Straaldyrene –. *Oversigt over det Kongelige Danske Videnskabernes Selska bs forhandlinger,* 77, 75–158.

Lütken, C.F. & Mortensen, T. (1899) Reports on an exploration off the west coasts of Mexico, Central and Southern America and off the Galapagos Islands. XXV. The Ophiuridae. *Memoirs of the Museum of Comparative Zoology,* 23, 97–208.

Lyman, T. (1860) Descriptions of new Ophiuridae, belonging to the Smithsonian Institution and to the Museum of Comparative Zoölogy at Cambridge. *Proceedings of the Boston Society of Natural History, 1859 to 61* 7, 193–204.

CHAPTER XII:

PHYLUM CHORDATA–SUBPHYLUM, VERTEBRATA–CLASS, PISCES– "THE BONY NIECES SEND MESSAGES FROM THE DEEP"

"Any suppression which undermines and destroys that very foundation on which scientific methodology and research was erected, evolutionist or otherwise, cannot and must not be allowed to flourish ... It is a confrontation between scientific objectivity and ingrained prejudice – between logic and emotion – between fact and fiction ... In the final analysis, objective scientific logic has to prevail – no matter what the final result is – no matter how many time-honoured idols have to be discarded in the process ... After all, it is not the duty of science to defend the theory of evolution and stick by it to the bitter end -no matter what illogical and unsupported conclusions it offers ... If in the process of impartial scientific logic, they find that creation by outside intelligence is the solution to our quandary, then Let's cut the umbilical cord that ties us down to Darwin for such a long time. It is choking us and holding us back ... Every single concept advanced by the theory of evolution (and amended thereafter) is imaginary as it is not supported by the scientifically established probability concepts. Darwin was wrong... The theory of evolution may be the worst mistake made in science." I L Cohen

The group of animals referred, collectively, as fish, inhabits all oceans, the deep sea and all bodies of fresh water. It is generally believed that they evolve about 460 million years ago in the Ordovician period. These groups of animals have been grouped into five classes with over 69,000 species. For this study, the osteichthyes, or bony fish was utilized and do have well developed bony scales and ossified bony skeletons. Their lungs, or swim bladders are well developed and the gill pouches open into a common chamber covered by an operculum. This class of fish does have over 22,000 species (see Figure 12.1).

Figure 12.1: Picture of Fishes.

Body Structure and Movement

The body of the fish is covered by bony scales which are divided into the head, the main body, and the tail. The head carries two eyes situated anterolaterally. Between these lie the noses. On the surface of the body, there are dorsal, median, paired pectoral and pelvic fins. The body of the fish is highly streamlined and enables it to move effortlessly through the water. There are integumentary mucous glands that secrete mucous on the surface of the fish and this reduces friction between the body surface and water during movement. The mucous, also, prevent unnecessary gas exchanges and protect the body against ectoparasites. The muscles of the trunk contain segmented myomeres, which contract to allow the fish to swim by lateral undulations. These undulations pass posteriorly from the trunk to the tail, and do allow fast swimming. To maintain stability and remain buoyant at a particular depth, the fish uses the lift of the pectoral fins and the lateral movements of the tail. Further, buoyancy is perfected by the swim bladder or lungs, which could be filled with gas to help maintain a body density similar to the surrounding water. The dorsal fins, the paired pectoral fins, the pelvic fins, and the median fins allow the fish to swim rapidly without rolling from side to side, and also prevent yawing. The paired fins do control the fish braking, turning and changing of its depth.

Nervous System

The sensory organs of fish are adapted for receiving stimuli in water. The eyes of the fish do not need tear glands and moveable lids, since they are already surrounded with water. The nose is only used as an olfactory organ and not for gas exchanges. This organ is highly developed since fish need this smell organ to identify their surroundings and find food and mates. The animals have developed an inner ear which allows them to detect pressure waves. There are also lateral line systems that allow the fish to detect low frequency vibrations, water movements and the ability to detect changes in pressure levels. The development of the nerves, the notochord and the neural system are still primitive as compared to other vertebrates.

Circulatory, Respiratory and Excretory System

The circulatory system in the fish is well developed. The heart is located ventrally in the anterior part of the trunk. Blood from capillaries in the trunk's tissues and from the kidney empty into the cutaneous vein, and the posterior cardinal vein respectively. The capillaries then move blood to the heart. The heart pumps the blood through the gills where they are oxygenated. From the gills, the blood passes to the dorsal and cutaneous aorta, which supplies blood directly under low pressures to tissues and organs of the abdomen and trunk. Gas exchanges occur at the gills. Water is taken in through the mouth and expelled under pressure through the gills. Oxygen taken in from the water is absorbed directly straight into the blood. The gills also help in excretion. Besides expelling nitrogenous metabolic wasters by diffusion through the surrounding water it helps expel sodium, potassium and calcium ions by special glands, or salt-excreting cell on the gills. The functions of water balance and excretion are still carried out by the kidneys, or the opisthonephros.

Nutrition

Fish do ingest food through the mouth by swallowing or using the jaws to seize the prey. The mouth does have teeth that allow the food to be crushed before passing to the pharynx and then the stomach. Digestion starts from the mouth. The food is further digested by secretions of enzymes from glands around stomach. The intestine, the pyloric caese, the liver and spleen aid in the digestive process. The liver also helps in storage of digested food and synthesis of glycogen. The digested food is normally absorbed through the intestinal wall and the undigested food matter is expelled through the anus opening just below the tail, or the anal fin (see Figure 12.2).

Figure 12.2: Picture of Fishes spawning.

Reproduction

The fish used in the study is oviparous and fertilization is external. The gonads are located in the posterior region of the trunk. The urogenital tract opens ventrally close to the anal fins. The sexes are separate. During copulation, the eggs are laid in the open water and the males pass out sperm that immediately fertilizes these eggs. Development is rapid and young fish normally develop from these zygotes.

Experiment with Mutagens

The common perch was used in the study and cultures containing young adult, matured adult and germ cells were exposed to mutagens of low, medium and high intensity. Some of the cultures were exposed to one of the mutagens only, and in combination of two and three mutagens. The duration of the cultures ranged from three months, six months, and one year to three years. Under low doses of x-rays, ultraviolet rays and radioactive solutions, it was observed that the perch continued normal metabolic activity. They seem unaffected by the mutagens. The reproductive activity was unaffected and no mutant variety was observed. Under medium intensity of the mutagens, there was a slowdown in metabolic activities. The animals became bloated, which was adduced to the failing of the functioning of the kidneys, or, an attempt by the organism to reduce the effect of the mutagens by retaining more fluid. The production of young fish from the fertilized eggs was drastically reduced. Both the adults and the young, or newly hatched fish, did not survive past the first 35 days. There were no observed mutant variations. Under high intensity of the mutagens, the adult perch and the newly hatched perch did not survive pass the first ten days.

Genetic Mutations

Cells from each of the cultures with exposure to the different intensities of the mutagens were examined to see if there were any genotypical changes. Observations revealed that there were no mutations taking place. The chromosome number remained unchanged. The genome was constant. The gene size and gene number remained the same. A new organelle associated with the chromosomes termed the MasterCodon was observed which controls genotypic and phenotypic changes.

Conclusion

The study did not reveal any significant mutations. The perch resist any attempt to mutate. The organelle responsible for this control is the MasterCodon. The MasterCodon seems to exert influence on the cell and organism is such a way as to prevent any attempt at mutations. This organelle enables the maintenance of cellular constancy, uniqueness of the cell, and ultimately ensuring that the organism does not change its subphylum from generation to generation. It, therefore, follows that the MasterCodon will kill the organism or shut down the cell's system rather than allow it to mutate into a different class of animals.

Suggested Reading

P.L. Angermeier, M.R. Winston. Characterizing fish community diversity across Virginia landscapes: prerequisite for conservation
Ecol. Appl., 9 (1) (1999), pp. 335-349

M.C. Barber. Bioaccumulation and Aquatic System Simulator (BASS) User's Manual Version 2.3
U.S. Environmental Protection Agency, Washington, D.C. (2012)
Report No.: EPA/600/R-01/035

M.C. Barber, B. Rashleigh, M. Cyterski. Forecasting fish biomasses, densities, productions, and bioaccumulation potentials of mid-Atlantic wadeable streams
Integrated Environ. Assess. Manag., 12 (1) (2015), pp. 146-159

M. Barbour, J. Gerritsen, B. Snyder, J. Stribling. Rapid Bioassessment Protocols for Use in Streams and Wadable Rivers: Periphyton, Benthic Invertebrates and Fish
U.S. Environmental Protection Agency, Office of Water, Washington, D.C. (1999)
Report No.: EPA 841/B-99/002

C. Baxter, F. Hauer. Geomorphology, hyporheic exchange, and selection of spawning habitat by bull trout (*Salvelinus confluentus*)
Can. J. Fish. Aquat. Sci., 57 (7) (2000), pp. 1470-1481

A. Bertolo, P. Magnan. Spatial and environmental correlates of fish community structure in Canadian shield lakes
Can. J. Fish. Aquat. Sci., 63 (12) (2006), pp. 2780-2792

T. Bohlin, C. Dellefors, U. Faremo, A. Johlander. The energetic equivalence hypothesis and the relation between population density and body size in stream-living salmonids
Am. Nat., 143 (3) (1994), pp. 478-493

W. Boicourt, C. Gallegos, L. Harding Jr., E. Houde, M. Mallonee, C. McClain, *et al.*
Trophic Indicators of Ecosystem Health in Chesapeake Bay
U.S. Environmental Protection Agency, Washington, D.C. (2004)
Report No.: R828677C002

T.O. Brenden, L. Wang, P.W. Seelbach. A river valley segment classification of Michigan streams based on fish and physical attributes
Trans. Am. Fish. Soc., 137 (6) (2008), pp. 1621-1636

C. Carbone, J.L. Gittleman. A common rule for the scaling of carnivore density
Science, 295 (5563) (2002), p. 2273

D.M. Carlisle, J. Falcone, M.R. Meador. Predicting the biological condition of streams: use of geospatial indicators of natural and anthropogenic characteristics of watersheds
Environ. Monit. Assess., 151 (1) (2009), pp. 143-160

T. Chen, C. Guestrin (Eds.), 22nd ACM SIGKDD. International Conference on Knowledge Discovery and Data Mining, ACM, San Francisco, CA, USA. New York, NY, USA (2016)

M.J. Cyterski, M.C. Barber. Identification and prediction of fish assemblages in streams of the mid-Atlantic Highlands, USA
Trans. Am. Fish. Soc., 135 (2006), pp. 40-48

D. Dauwalter, D. Splinter, W. Fisher, R. Marston. Biogeography, ecoregions, and geomorphology affect fish species composition in streams of eastern Oklahoma, USA
Environ. Biol. Fish., 82 (3) (2008), pp. 237-249

W. Davis, B. Snyder, J. Stribling, C. Stoughton. Summary of State Biological Assessment Programs for Streams and Rivers
U. S. Environmental Protection Agency Office of Planning, Policy, and Evaluation, Washington, DC (1996)
Report No.: EPA 230/R-96/007

D. Duplisea, M. Castonguay. Comparison and utility of different size-based metrics of fish communities for detecting fishery impacts
Can. J. Fish. Aquat. Sci., 63 (4) (2006), pp. 810-820

T. Erős, P. Sály, P. Takács, C. Higgins, P. Bíró, D. Schmera. Quantifying temporal variability in the metacommunity structure of stream fishes: the influence of non-native species and environmental drivers.
Hydrobiologia, 722 (1) (2014), pp. 31-43

M.R. Falcy, J.L. McCormick, S.A. Miller. Proxies in practice: calibration and validation of multiple indices of animal abundance.
J. Fish Wildlife Manag., 7 (1) (2016), pp. 117-128

K. Fausch, J. Lyons, J. Karr, P. Angermeier. Fish communities as indicators of environmental degradation.
S. Adams (Ed.), Biological Indicators of Stress in Fish. Symposium 8, American Fisheries Society, Bethesda, MD (1990), pp. 123-144

A.H. Fielding. Machine Learning Methods for Ecological Applications.
Kluwer Academic Publishers, Boston, MA (1999)

B. Fransen, S. Duke, G. McWethy, J. Walter, R. Bilby. A logistic regression model for predicting the upstream extent of fish occurrence based on geographical information systems data.
N. Am. J. Fish. Manag., 26 (4) (2006), pp. 960-975

M.C. Freeman, M.K. Crawford, J.C. Barrett, D.E. Facey, M.G. Flood, J. Hill, *et al.* Fish assemblage stability in a southern appalachian stream.
Can. J. Fish. Aquat. Sci., 45 (11) (1988), pp. 1949-1958

G.D. Grossman, J.F. Dowd, M. Crawford. Assemblage stability in stream fishes: a review. Environ. Manag., 14 (5) (1990), pp. 661-671

B. Han, M. Straškraba. Size dependence of biomass spectra and population density I. The effects of size scales and size intervals.
J. Theor. Biol., 191 (3) (1998), pp. 259-265

A. Herlihy, D. Larsen, S. Paulsen, S. Urquhart, B. Rosenbaum. Designing a spatially balanced, randomized site selection process for regional stream surveys: the EMAP mid-Atlantic pilot study.
Environ. Monit. Assess., 63 (1) (2000), pp. 95-113

M.H. Weber, S.G. Leibowitz, A.R. Olsen, D.J. Thornbrugh The stream-catchment (StreamCat) dataset: a database of watershed metrics for the conterminous United States
JAWRA J. Am. Water Resour. Assoc., 52 (1) (2016), pp. 120-128

CHAPTER XIII:

PHYLUM, CHORDATA—SUBPHYLUM, VERTEBRATA, CLASS, AMPHIBHIA, "OUR STURDY, GRUMPY AUNTIES SEND THEIR LOVE!"

"The theory of Evolution ... will be one of the great jokes in the history books of the future. Posterity will marvel that so flimsy and dubious an hypothesis could be accepted with the incredible credulity it has." (Malcolm Muggeridge, well-known philosopher) "Scientists who go about teaching that Evolution is a fact of life are great con men, and the story they are telling may be the greatest hoax ever. In explaining Evolution we do not have one iota of fact." Dr T N Tahmisian

The group of animals belonging to the class Amphibia was believed to have evolved about 400 million years ago in the Devonian period. There are over 4,000 existing species of this class and the most common are the frogs, salamanders and caecilians. For this study, the common clawed frog, Xenopus Levis, was chosen (see Figure 13.1).

Figure 13.1: Figure of a Frog.

Body Structure and Movement

The animals belonging to this group have skin with abundant mucous glands, keratin, and poison gland. The keratinized epidermis helps prevent water loss from, and abrasiveness of the frog's skin. The outer part the dermis does have pigment cells, or, chromophores, which protect the animals against ultraviolet rays. The mucous glands in dermis secrete viscous substance that helps keep the animal moist, and also acts as a defense mechanism. The dermis is also highly vascularized and enables gas exchange. In addition, the dermis allows the organism to loose water rapidly during excretion process. A highly developed skeletal system exists in the frogs as an adaptation to terrestrial living and for walking and hopping on land. Also, it has a developed and strong musculature associated with the skeletal system that is used for raising the head for feeding, and for raising the body by the limbs during movement. The limbs are supported by strong girdles, and the fore limbs are shorter than the hind limbs. The hind limbs are modified to help the frog push for leaping, jumping and hopping.

Nervous System

At the dorsa-lateral section of the head are pair of eyes that are protected and cleansed by the eyelids and the tear glands. The cornea has been modified to allow for more refraction of light. At the anterior end of the head is a pair of nostrils which are used in detecting smell from chemicals and other substances. There are is pair of tympanic membranes just behind the eyes that are used for detecting vibrations, airborne sound waves and mating calls. The brain is not highly developed in the frogs and messages are relayed to and from the brain through a spinal cord which is protected by the vertebra.

Circulatory System, Excretion and Respiration

The heart is well developed in frogs with a right and left atria and a single ventricle. The blood from the head, other parts of the body, and the cutaneous artery is rich in oxygen from the skin. This oxygenated blood enters the head through the right atrium. Oxygen- rich blood from the lungs enters heart through the left atrium. Both of atria's entries of blood to the heart are prevented from mixing in the ventricle. Blood rich in oxygen is sent to the head; and mixed blood is sent to other parts of the body. Oxygen-depleted blood is sent to the lungs and by cutaneous vein to the skin. Respiration or, gas exchanges takes place through the lungs and the oxygenated blood in the lungs is carried via the pulmo-cutaneous arch to the heart. Excretion process is carried out by the kidneys. Frogs do lose a lot of water, and since their water preservation ability is still rudimentary, the nitrogenous waste excreted through the kidney is highly concentrated urea and requires more water intake to flush it out.

Nutrition

The frogs are ectotherms in that their body temperatures fluctuate with the environmental temperature. They prefer habitats with cool and moist surroundings. Since they have low metabolic rate, their need for food and oxygen is relatively low. Feeding is carried out by means of movements of the head, jaws-hyoid apparatus, and a muscular, moist, and sticky tongue. The tongue could be extended out of the mouth to capture invertebrate preys including flying insects. Theses preys are then pulled back into the mouth. The prey is held back by the teeth. Digestion starts immediately in the mouth by secretions from buccal and pharyngeal glands. An enzyme called chitinase is produced by the gastric glands and pancreas which empties into the stomach and is used to digest the chitinous cuticle of insects. Digestion is completed in the small intestine and absorption of the digested food into the blood system takes place. The undigested food is sent through the large intestine for more water absorption before the food waste is sent out through the anus.

Reproduction

Reproduction in frogs does occur close to aquatic environment. More of the mating processes occur around ponds or other fresh water during spring time. The male makes mating noise to attract the female during mating; a process referred to as amplexus. The male sits astride the female and grasp it, and as the eggs are being laid in the water, the sperms are shed on them and fertilization take place immediately. The fertilized eggs develop into the larval stage called the tadpoles. Initially, the larval stage does have external gills which are shed after metamorphosis. Then, new internal gills are developed. During this stage, the tadpole develops the hind limbs and is usually herbivorous during feeding. After more metamorphosis, they gradually shed their aquatic features and develop into the adult stage with terrestrial features (see Figure 13.2).

Figure 13.2: Picture of Frogs' mating behavior.

Experiment with Mutagens

Various cultures of frogs were exposed to three different levels of intensities of mutagens. The cultures contained the adult stage, the larval stage and the germ cells. Exposures of these cultures were subjected to ultraviolet rays. X –rays and radioactive elements. The cultures were exposed to the mutagens individually, and in combinations of two and three of the mutagens. The duration was for periods of one month, three months, six months, one year, and three years. The study involved the observation of the organism as to ascertain the effect of the mutagens on the animal. Under low intensity of the mutagens, it was observed that the animal continued normal metabolic activity. Reproduction was normal. Occasionally, the animal becomes frequently aestivated. There were no observed mutations. With exposure to medium intensity of the mutagens, the frog seems to react more adversely to cultures with x- rays and radioactive elements. Metabolic activity was greatly reduced. The period of estivation was longer. Reproductive activities were greatly reduced. The animal took in, or absorbs more fluid. The larval forms had shorter bodies and undeveloped hind limbs. They did not survive into the adult stages. The number of fertilized eggs developing into the larval forms was greatly reduced. There were no observed mutations. Under high intensity of the mutagens, the adult stage and the larval stage did not survive past the fifth days. The fertilized eggs did not develop, or survive. Few mutant varieties were observed where larvae had no gills, no internal organs, and no mouths.

Genetic Mutations

Cells from each of the cultures were investigated to observe the effect of the mutagens on nuclei content with the intent of observing any genetic mutations. It appears the change observed were only phonotypical. The chromosomes were not affected. The genome of 3.1×10^0bp was unchanged. The mutations observed in the developed tadpoles with abnormalities were due to genetic base shift resulting in the coding of different proteins, which will normally be produced for the specific cells, or tissues of the affected parts. A new organelle associated with the chromosome, which was termed the MasterCodon was observed to be the controlling factor preventing genotypic mutations resulting in forced death.

Conclusion

From the study, it was observed that the frog did resist any attempt to mutate. The effect of the mutagens was detrimental, or, lethal to the organism. Rather than mutate, the organism had its whole system shutdown. The organelle exerting this influence is the MasterCodon. It acts as control over the chromosomes and the genes. It maintains the constancy and uniqueness of the organism to adapt to favorable conditions, or adverse conditions; however, prevent genotypical mutations and ensure that the organism's uniqueness is maintained. Rather than allow mutations, the MasterCodon will shut down the cellular system and forced the organism to die.

Suggested Reading

ABRAVAYA, J.P. & JACKSON, J.F. 1978. Reproduction in *Macrogenioglottus alipioi* Carvalho (Anura, Leptodactylidae). Nat. Hist. Mus. Los. Ang. Cty. Contrib. Sci. 298:1-9. [Links]

ABREU, R.O., JUNCÁ, F.A., SOUZA, I.C.A. & NAPOLI, M.F. 2015. The tadpole of *Dendropsophus branneri* (Cochran, 1948) (Amphibia, Anura, Hylidae). Zootaxa 3946(2):296-300. [Links]

ABREU, R.O., NAPOLI, M.F., CAMARDELLI, M. & FONSECA, P.M. 2013. The tadpole of *Dendropsophus haddadi* (Amphibia, Anura, Hylidae): additions on morphological traits and comparisons with tadpoles of the *D. decipiens* and *D. microcephalus* species groups. Sitientibus Série Ciências Biológicas 13:1-4. [Links]

ABREU, R.O., NAPOLI, M.F., TREVISAN, C.C., CAMARDELLI, M., DÓRIA, T.A.F. & SILVA, L.M. 2015. The tadpole of *Scinax melanodactylus* (Lourenço, Luna & Pombal Jr, 2014) (Amphibia, Anura, Hylidae). Zootaxa 3981(3):430-436. [Links]

ALMEIDA, J.P.F.A., NASCIMENTO, F.A.C., TORQUATO, S., LISBOA, B.S., TIBÚRCIO, I.C.S., PALMEIRA, C.N.S., LIMA, M.G. & MOTT, T. 2016. Amphibians of Alagoas State, northeastern Brazil. Herpetol. Notes 9:123-140. [Links]

ALTIG, R. & MCDIARMID, R. W. 1999. Body plan: Development and morphology. In Tadpoles: The Biology of Anuran Larvae (R.W. McDiarmid & R. Altig, eds). University of Chicago Press, Chicago, p.24-50. [Links]

ALTIG, R. 1970. A key to the tadpoles of the continental United States and Canada. Herpetologica 26(2):180-207. [Links]

ALVES, A.C.R., GOMES, R. & CARVALHO, S.P. 2004. Description of the tadpole of *Scinax auratus* (Wied-Neuwied) (Anura, Hylidae). Rev. Bras. Zool. 21(2):315-317. [Links]

AMPHIBIAWEB. 2020. https://amphibiaweb.org. University of California, Berkeley, CA, USA (last access in 09/Apr/2020). [Links]

ANDRADE, G.V., ETEROVICK, P.C., ROSSA-FERES, D.C. & SCHIESARI, L. 2007. Estudos sobre girinos no Brasil: histórico, situação atual e perspectivas. In Herpetologia no Brasil II (L.B. Nascimento & M.E. Oliveira, orgs). Soc. Bras. Herpetol., Belo Horizonte, p.127-145. [Links]

ASSIS, J.S. 2000. Biogeografia e conservação da biodiversidade–projeções para Alagoas. Edições Catavento, Maceió. [Links]

BARBOSA, V.N., PEREIRA, E.N., & MARANHÃO, E.S. 2017. Anfíbios da Estação Ecológica de Caetés Paulista, Pernambuco–Atualização da lista de espécies. Rev. Ciên. Amb. 11(2):39-49.

Rev. Bras. Biol. 22(4):391-399. [Links]

BOKERMANN, C.A. 1963. Girinos de anfíbios brasileiros–I (Amphibia–Salientia). An. Acad. Bras. Ciênc. 35(3):465-474. [Links]

BOKERMANN, W.C.A. 1973. Duas novas espécies de Sphaenorhynchus da Bahia (Anura, Hylidae). Rev. Bras. Biol. 33(4):589-594. [Links]

BOULENGER, G.A. 1892. Description of new reptiles and batrachians from the Loo Choo islands. Ann. Mag. Nat. Hist. 13(18):302-304. [Links]

BREDER, J.C.M. 1946. Amphibians and reptiles of the rio Chucunaque drainage, Darien, Panama, with notes on their life histories and habits. Bull. Am. Museum Nat. Hist. 86(8):375-436. [Links]

CAMPOS, T.F., LIMA, M.G., NASCIMENTO, F.A.C. & SANTOS, E.M. 2014. Larval morphology and advertisement call of Phyllodytes acuminatus Bokermann, 1966 (Anura: Hylidae) from Northeastern Brazil. Zootaxa 3779(1):93-100. [Links]

CARAMASCHI, U. 1979. O girino de Odontophrynus carvalhoi Savage & Cei, 1965 (Amphibia, Anura, Ceratophrydidae). Rev. Bras. Biol. 39(1):169-171. [Links]

CARNAVAL, C.A.O.Q. & PEIXOTO, O.L. 2004. A new species of Hyla from northeastern Brazil (Amphibia, Anura, Hylidae). Herpetologica 60(3):387-395. [Links]

CARNEIRO, M.C.L., MAGALHÃES, P.S. & JUNCÁ, F.A. 2004. Descrição do girino e vocalização de Scinax pachycrus (Miranda-Ribeiro, 1937) (Amphibia, Anura, Hylidae). Arq. do Mus. Nac. 62(3):241-246. [Links]

CARVALHO-E-SILVA, S.P., CARVALHO-E-SILVA, A.M.P.T. & IZECKSOHN, E. 2003. Nova espécie de *Hyla laurenti* do grupo de *H. microcephala* Cope (Amphibia, Anura, Hylidae) do nordeste do Brasil. Rev. Bras. Zool. 20(3):553-558. [Links]

CARVALHO-E-SILVA, S.P., CORREA PINTO, A.L. & CARVALHO-E-SILVA, A.M.P.T. 2002. Aspectos da reprodução, da vocalização e da larva *Phrynohyas mesophaea* (Amphibia, Anura, Hylidae). Rev. Aquarium 15:19-24. [Links]

CASAL, F.C. & JUNCÁ, F.A. 2008. Girino e canto de anúncio de *Hypsiboas crepitans* (Amphibia: Anura: Hylidae) do estado da Bahia, Brasil, e considerações taxonômicas. Bol. do Mus. Para. Emilio Goeldi 3(3):217-224. [Links]

CASCON, P. & PEIXOTO, O.L. 1985. Observações sobre a larva de *Leptodactylus troglodytes* A. Lutz, 1926 (Amphibia, Anura, Leptodactylidae). Rev. Bras. Biol. 45:361-364. [Links]

CEI, J. M. 1980. Amphibians of Argentina. Monit. Zool. Ital., N.S., Monografia. [Links]

COSTA, E.F., NASCIMENTO, F.A.C., JÚNIOR, M. M. & SANTOS, E.M. 2017. Aspectos de vida de *Odontophrynus carvalhoi* Savage & Cei, 1965 (Amphibia, Anura, Odontophrynidae) em um brejo de altitude no Nordeste brasileiro. Bol. Mus. Biol. Mello Leitão 39(2):95-115. [Links]

CRUZ, C.A.G. 1982. Conceituação de grupos de espécie de Phyllomedusinae brasileiras com base em caracteres larvários (Amphibia, Anura, Hylidae). Arq. Univ. Fed. Rur. Rio de J. 5(2):147-171. [Links]

CRUZ, C.A.G. & PEIXOTO, O.L. 1982. Sobre a biologia de *Atelopus pernambucencis* Bokermann, 1962 (Amphibia, Anura, Bufonidae). Rev. Bras. Biol. 42(3):627-629. [Links]

CRUZ, C.A.G., NUNES, I., & JUNCÁ, F.A. 2012. Redescription of *Proceratophrys cristiceps* (Müller, 1883) (Amphibia, Anura, Odontophrynidae), with description of two new species without eyelid appendages from northeastern Brazil. S. Am. J. Herpetol. 7: 110-122. [Links]

DE SÁ, R.O., GRANT, T., CAMARGO, A., HEYER, W.R., PONSSA, M.L. & STANLEY, E. 2014. Systematics of the neotropical genus *Leptodactylus* Fitzinger, 1826 (Anura: Leptodactylidae): phylogeny, the relevance of non-molecular evidence, and species accounts. S. Am. J. Herpetol. 9(s1): S1-S128. [Links]

DIXON, JR. & STATON, M.A. 1976. Some aspects of the biology of *Leptodactylus macrosternum* Miranda-Ribeiro (Anura: Leptodactylidae) of the Venezuelan Llanos. Herpetologica 322:227-232. [Links]

DUBEUX, M.J.M., DA-SILVA, G.R.S., NASCIMENTO, F.A.C., GONÇALVES, U., & MOTT, T. 2019. Síntese histórica e avanços no conhecimento de girinos (Amphibia: Anura) no estado de Alagoas, nordeste do Brasil. Rev. Nordest. Zool. 12(1):18-52. [Links]

DUBEUX, M.J.M., SILVA, T., MOTT, T & NASCIMENTO, F.A.C. 2020a. Redescription of the tadpole of *Leptodactylus natalensis* Lutz (Anura: Leptodactylidae), an inhabitant of the Brazilian Atlantic Forest. Zootaxa 4732:346-350. [Links]

DUBEUX, M.J.M., GONÇALVES, U., NASCIMENTO, F.A.C. & MOTT, T. 2020b. Anuran amphibians of a protected area in the Northern Atlantic Forest with comments on topotypic and endangered populations. Herpetol. Notes 13:61-74.

DUELLMAN, W.E. 1978. The biology of an equatorial herpetofauna in Amazonian Ecuador. Misc. Publ. Mus. Nat. Hist. Univ. Kansas 65:1-352. [Links]

DUELLMAN, W.E. & TRUEB, T. 1994. Biology of Amphibia. 2ª ed. The Johns Hopkins University Press, Baltimore. [Links]

DUELLMAN, W.E. & TRUEB, L. 2015. Marsupial frogs, *Gastrotheca* and allied genera. Johns Hopkins University Press, Baltimore. [Links]

FABREZI, M. & VERA, R. 1997. Caracterización morfológica de larvas de anuros del Noroeste Argentino. Cuad. Herpetol. 11(1-2):37-49. [Links]

FAIVOVICH, J. 2002. A cladistic analysis of *Scinax* (Anura: Hylidae). Cladistics 18(4):367-393. [Links]

FAIVOVICH, J., HADDAD, C. F., GARCIA, P. C., FROST, D. R., CAMPBELL, J. A., & WHEELER, W. C. 2005. Systematic review of the frog family Hylidae, with special reference to Hylinae: phylogenetic analysis and taxonomic revision. Bull. Am. Mus. Nat. Hist. 2005(294):1-240. [Links]

FATORELLI, P., NOGUEIRA-COSTA, P. & ROCHA, C.F.D. 2017. Characterization of tadpoles of the southward portion (oceanic face) of Ilha Grande, Rio de Janeiro, Brazil, with a proposal for identification key. North-West. J. Zool. 4(2):171-184. [Links]

FROST, D.R. 2020. Amphibian species of the world, version 6.0. American Museum of Natural History, New York, USA Online Reference. http://research. amnh.org/herpetology/amphibia/index.php (last access in 09/Apr/2020). [Links]

GOMES, M.R. & PEIXOTO, O.L. 1991a. Larvas de *Hyla* do grupo *"leucophyllata"* com a descrição de *H. elegans* Wied, 1824 e notas sobre a variação do pradrão de colorido do adulto nesta espécie (Anura, Hylidae). Rev. Bras. Biol. 51:257-262. [Links]

GOMES, M.R. & PEIXOTO, O.L. 1991b. Considerações sobre os girinos de *Hyla senicula* (Cope, 1868) e *Hyla soaresi* (Caramaschi e Jim, 1983) (Amphibia, Anura, Hylidae). Acta Biol. Leopold. 13:5-18. [Links]

GOMES, M. R., ALVES, A.C.R. & PEIXOTO, O.L. 2014. O girino de *Scinax nebulosus* (Amphibia, Anura, Hylidae). Iheringia. Série Zool. 104(2):184-188. [Links]

GOSNER, K.L. 1960. A simplified table for staging anuran embryos and larvae with notes on identification. Herpetologica 16(3):183-190. [Links]

GROSJEAN, S. 2005. The choice of external morphological characters and developmental stages for tadpole-based anuran taxonomy: a case study in *Rana* (*Sylvirana*) *nigrovittata* (Blyth, 1855)(Amphibia, Anura, Ranidae). Contrib. Zool. 74(1-2):61-76. [Links]

HAAS, A., WOLTER, J., HERTWIG, S.T. & DAS, I. 2009. Larval morphologies of three species of stream toads, genus *Ansonia* (Amphibia: Bufonidae) from east Malaysia (Borneo), with a key to known Bornean *Ansonia* tadpoles. Zootaxa 2302(1):21-18. [Links]

HEDGES, S.B., DUELLMAN, W.E. & HEINICKE, M.P. 2008. New World direct-developing frogs (Anura: Terrarana): molecular phylogeny, classification, biogeography, and conservation. Zootaxa 1737(1):1-182. [Links]

HERO, J.M.. 1990. An illustrated key to tadpoles occurring in the Central Amazon rainforest, Manaus, Amazonas, Brazil. Amazoniana 11(2):201-262. [Links]

HEYER, W.R. 1978. Systematics of the fuscus group of the frog genus *Leptodactylus* (Amphibia, Leptodactylidae). Sei. Natur. Hist. Mus. Los Angeles County. 29:1-85. [Links]

HEYER, W.R., RAND, A.S., CRUZ, C.A.G., PEIXOTO, O.L. & NELSON, C.E. 1990. Frogs of Boracéia. Arq. Zool. 31(4):231-410. [Links]

ICMBIO. 2018. Fauna brasileira ameaçada de extinção. In: Fauna Brasileira Ameaçada de Extinção. Fundação Biodiversitas para a Conservação da Diversidade Biológica. Brasília, Distrito Federal. [Links]

IUCN. 2020. The IUCN Red List of Threatened Species. Version 2020-1, International Union for Conservation of Nature and Natural Resources. http://www. iucnredlist.org (last access in 09/Apr/2020). [Links]

JUNCÁ, F.A., CARNEIRO, M.C.L. & RODRIGUES, N.N. 2008. Is a dwarf population of *Corythomantis greeningi* Boulenger, 1896 (Anura, Hylidae) a new species? Zootaxa 1686 (56):48-56. [Links]

KENNY, J.S. 1968. The Amphibia of Trindad. Stu. Fauna Curaçao other Carib. Isl. 29:1-78. [Links]

KOKUBUM, M.N.C. & DE SOUSA, M.B. 2008. Reproductive ecology of *Leptodactylus* aff *hylaedactylus* (Anura, Leptodactylidae) from an open area in northern Brazil. S. Am. J. Herpetol. 3:15-21. [Links]

KOKUBUM, M.N.C. & MACIEL, N.M. 2009. Reproductive biology of the Brazilian sibilator frog *Leptodactylus troglodytes*. Herpetol. J. 19:119-126. [Links]

KOLENC, F., BORTEIRO, C., ALCALDE, L., BALDO, D., CARDOZO, D. & FAIVOVICH, J. 2008. Comparative larval morphology of eight species. Zootaxa 1927(1):1-66. [Links]

LANTYER-SILVA, A. S., MATOS, M. A., GOGLIATH, M., MARCIANO-JR, E. & NICOLA, P. A. 2016. New records of *Pseudopaludicola pocoto* Magalhães, Loebmann, Kokubum, Haddad & Garda, 2014 (Amphibia: Anura: Leptodactylidae) in the Caatinga Biome, Brazil. CheckList 12(6):1-4. [Links]

LAUFER, G., PEREYRA, L.C., AKMENTINS, M.S., & BORTEIRO, C. 2013. A comment on the oral dermal flaps of *Elachistocleis* Parker, 1927 (Anura: Microhylidae) larvae. Zootaxa 3710(5): 498-500. [Links]

LAVILLA, E.O. 1990. The tadpole of *Hyla nana* (Anura: Hylidae). J. Herpetol. 24(2):207-209. [Links]

LAVILLA, E.O. 1992. The tadpole of *Dermatonotus muelleri* (Anura: Microhylidae). Bolletino del Mus. Reg. di Sci. Nat. Torino 10(1):62-71. [Links]

LEITE-FILHO, E., DE-OLIVEIRA, F.A., ELOI, F.J., LIBERAL, C.N., LOPES, A.O., & MESQUITA, D.O. 2017. Evolutionary and ecological factors influencing an anuran community structure in an Atlantic Rainforest urban Fragment. Copeia 105(1):64-74.

Environ. 31(1):17-26. [Links]

LISBOA, B., SANTOS, W. F. S., TORQUATO, S., GUARNIERI, M. C., & MOTT, T. 2019. A new state record of the glassfrog *Vitreorana baliomma* (Anura: Centrolenidae), with notes on its reproductive biology. Herpet. Notes 12:957-960. [Links]

LISBOA, B.S., NASCIMENTO, F.A.C. D. & SKUK, G.O. 2011. Redescription of the tadpole of *Macrogenioglottus alipioi* (Anura: Cycloramphidae), a rare and endemic species of the Brazilian Atlantic Forest. Zootaxa 3046(1):67-68. [Links]

LOEBMANN, D., ORRICO, V.G.D. & HADDAD, C.F.B. 2011. First record of *Adelophryne baturitensis* Hoogmoed, Borges & Cascon, 1994 for the state of Pernambuco, northeastern Brazil (Anura, Eleutherodactylidae, Phyzelaphryninae). Herpetol. Notes 4:75-77. [Links]

LUTZ, B. 1973. Brazilian species of *Hyla*. University of Texas Press, Austin. [Links]

LYNCH, J.D. 2006. The tadpoles of frogs and toads found in the lowlands of northern Colombia. Rev. la Acad. Colomb. Ciencias Exactas, Fis. y Nat. 30(116):443-457. [Links]

MACHADO, I.F. & MALTCHINK, L. 2007. Check-list da diversidade de anuros no Rio Grande do Sul (Brasil) e proposta de classificação para as formas larvais. Neotrop. Biol. Conserv. 2(2):101-116. [Links]

MACIEL, D.B., & NUNES, I. 2010. A new species of four-eyed frog genus *Pleurodema* Tschudi, 1838 (Anura: Leiuperidae) from the rock meadows of Espinhaço range, Brazil. Zootaxa 2640:53-61. [Links]

MAGALHÃES, F.M., SANTANA, D.J., NETO, A.M., & GARDA, A.A. 2012. The tadpole of *Elachistocleis* cesarii Miranda-Ribeiro, 1920 (Anura, Microhylidae). Zootaxa 3187: 54-56. [Links]

MAGALHÃES, F.M., GARDA, A. A., AMADO, T. F., & DE SÁ, R. O. 2013a. The tadpole of *Leptodactylus caatingae* Heyer & Juncá, 2003 (Anura: Leptodactylidae): External morphology, internal anatomy, and natural history. S. Am. J. Herpetol. 8(3):203-211.

CHAPTER XIV:

PHYLUM–CHIRDATA–SUBPHYLUM, VERTEBRATA–CLASS, REPTILIA. "THE SCALY UNCLES PAID A VISIT!"

"Evolution is a fairy tale for grown-ups. This theory has helped nothing in the progress of science. It is useless." Dr Louis Bounoure

The group of animals belonging to the class Reptilia was believed to have evolved about 300 million years ago during the Pennsylvanian period. They are amniotes and are known to have developed cleidoic eggs. They are ectothermic and more active than the amphibians. The external surface of the body is covered with horny scales or plates and has reduced cutaneous glands. The heart ventricle has become fully divided into two to prevent the mixing of oxygenated and de-oxygenated blood. Fertilization is internal and development is oviparous. Some animals belonging to the class Reptilia are snakes, lizard, turtles, tortoises, alligators, crocodiles, chameleons and the extinct dinosaurs. For this study, we decided to use the common lizard (see Figure 14.1).

Figure 14.1: Picture of a Lizard.

Body Structure and Movements

The outer surface of the animals or the epidermis is dry, horny and keratinous. The epidermal cells produce horny scales or plates which is an adaptation to terrestrial living. The dry scales protect the body against excessive water loss, abrasion and harmful radiation. The keratinized epidermal cells regularly shed old cells as new cells are developed. The skin has reduced glands which produce sticky substances that are used for mating recognition and for protection against predators. The skeleton is well-ossified and much stronger than in fish and the amphibians. The limbs have been modified for the development of claws that jut out of the toes. These help the animal to have a better grip while crawling on the ground. These claws are also used by some species for protection and feeding for grasping of prey. The muscles have become more complex and help in the adaptation to movement on land. These animals could run and walk. The hind limbs have been modified to allow for more thrust during running and walking. These modifications have resulted in the development of two sacral vertebrae, which help to transfer the thrust to the vertebral column.

Circulation, Respiration and Excretion

The heart does have chambers with the ventricles completely separated. The circulatory system has become well developed and forms the principal mode of transportation of digested food, oxygen and other metabolic substance to all parts of the body. Also, the blood capillaries have become well developed and carry blood to and from the tissues. Blood leaving the tissues and organs empties into the vena cavea, and then into the left systemic arch before moving to the heart. Blood that has been oxygenated from the lungs moves into the heart through the pulmonary vein and pumped under pressure through the right systemic arch, and, then, the common carotid to the other parts of the body. The lungs surface has been greatly increased as compared to the amphibians, since reptiles do not use their skin for respiration. The reptiles have aspiratory lungs, which allow the increase, or decrease in size and pressure of the body cavity. Further, during lungs' contraction, the trunk muscles and the elastic recoil of the lungs force the air in and out of the body cavity. Oxygen taken into the lungs is carried by hemoglobin in the blood to the heart where it is pumped to all other parts of the body. Excretion is carried out by a pair of metanephric kidneys which helps remove nitrogenous waste from the blood and expelled out of the body in crystals of uric acid. Since these are non-toxic, the reptiles do not need excess water to flush them out. Rather than expel much water, the water is re-absorbed by the kidney tubules, the urinary bladder and the cloaca.

Nutrition

The reptiles are ectothermic and this means that they could maintain their body heat by their exposure to the sun, or their environment. Unlike endothermic mammals that depends on oxidation of their food for maintaining their body heat; reptiles eat less food since their body heat maintenance depends on the environmental temperature. The high temperature maintained by the reptiles allows for much more rapid digestion. The lizard, which is the experimental animal, is a herbivorous animal. However, most species of the lizards are insectivorous. These species have developed adaptations which allow them to use long and sticky tongue for catching insects. Food digestion starts from the mouth with secretions from the salivary glands. The food is moved to the stomach, then the small intestine where gastric and pancreatic secretions help conclude digestion respectively. The large intestine does contain bacteria that help digest cellulose in plants that were ingested as food. Absorption of digested food usually takes place in the large intestines. Undigested food wastes are sent out as feces through the anus.

Reproduction

The sexes in reptiles are separate and fertilization is internal. They have developed copulatory organs for mating. They have also developed cleidoic eggs that allow the zygotes to feed and develop within the eggs, and at the same time being protected from predators. All metabolic activities take place within the eggs. The reptiles are generally oviparous and the internally fertilized eggs are laid in soil or sand or leaf mold or in decomposed plants. The zygotes develop into the young adult lizards within the eggs and hatch out by crawling out. The young lizards, then, gradually develop into the adult forms with their exposure to the environment (see Figure 14.2).

Figure 14.2: A picture of Reptiles' mating behavior.

Experiment with Mutagens

The study entails using the adult lizard, the developing embryo and the germ cells in different cultures with exposure to three levels of intensities of the mutagens. The mutagens were chosen because they occur freely in the environment in the form of sun ultraviolet rays and in combination with x-rays. Also, physical and chemical reagents were utilized to produce radioactive elements. The purpose of utilizing these mutagens in laboratory conditions was to simulate how the animals will react to the mutagens by creating spontaneous evolution or gradual evolution through transferable mutable adaptations. Under low intensity, the lizard continued normal feeding. Respiratory activity was normal. Locomotion or movements continued normally. The reproductive activity was not affected. There were no observed mutant varieties. Under medium intensity, the animal reacted negatively to the ultra-violet rays immediately. It hibernated more frequently and was only active when it was about to feed. Its reaction to the other mutagens was similar. Normal metabolic activities were greatly reduced. Reproductive activity gradually ceased. The continuous hibernation of the animal allowed it to survive longer. There were no observed mutant varieties. Under high intensity of the mutagens, it was observed that the adult and developing lizard did not survive past the 15th day. However, it was observed that germ cells exposed to the high intensity for six month, and then fertilized, did create few mutant varieties. The young lizards that emerged from the shell had no limbs; some had no internal organs for feeding and excretion. All the mutant varieties did not survive past the second day, despite having the shapes of small lizards.

Genetic Mutations

Excision of cells from each of the cultures were examined to observe any genetic mutation occurring. There were no changes in the chromosomal content of the animal. The genome of 2.2×10^9 BP remained constant. The gene size and total gene number also remained unchanged. However, in the mutant varieties, we did notice chromosomal reduction and breakages. A new organelle, which was termed the MasterCodon, was observed in association with the chromosomes. This organelle prevents genotypic mutations from taking place.

Conclusion

From the study, it was observed that the effect of the mutagens did not create any significant mutations. The few mutations that did occur were lethal to the organism. Mutations that resulted in adaptations to the environment appear phonotypical and do not generally affect the gene. However, some of these phonotypical changes did affect the genes in terms of sequence and not in terms of increase in numbers. In other words, adaptability mutations were results of genetic code changes which created different proteins from the DNA bases and did not affect the total number of genes present in the organism. The MasterCodon seems to allow point mutations and frame shift mutations that will allow adaptability of the organism to its environment. But, it discourages mutations that will change the total chromosomal content, the genome, and total gene number of the organism by shutting down the whole cell system. In other words, it protects the uniqueness of the cell, or the organism as a whole.

Suggested Reading

Murphy RW, Crawford AJ, Bauer AM, Che J, Donnellan SC, Fritz U, et al. Cold Code: the global initiative to DNA barcode amphibians and nonavian reptiles. Mol. Ecol. Res. 2013;13(2): 161–167.

Chovanec A, Grillitsch B. Gefährdeter Lebensraum » Kleingewässer «–Amphibien als Bioindikatoren. Österreichs Fischerei. 1994;47: 289–292.

Reichholf JH. Frösche als Bioindikatoren. Stapfia 47, zugleich Kataloge des O.Ö Landesmuseum N.F. 1996;107: 177–188.

Araújo MB. The coincidence of people and biodiversity in Europe. Glob Ecol. Biogeogr. 2003;12: 5–12.

Beebee TJC, Griffiths RA. The amphibian decline crisis: A watershed for conservation biology? Biol. Conserv. 2005;125: 271–285.

Schlaepfer MA, Hoover C, Dodd CK. Challenges in evaluating the impact of the trade in amphibians and reptiles on wild populations. BioScience. 2005;55: 256–264.

Hawlitschek O, Morinière J, Dunz A, Franzen M, Rödder D, Glaw F, et al. Comprehensive DNA barcoding of the herpetofauna of Germany. Mol. Ecol. Res. 2015;16: 242–253.

Heym A, Deichsel G, Hochkirch A, Veith M, Schulte U. Do introduced wall lizards (*Podarcis muralis*) cause niche shifts in a native sand lizard (*Lacerta agilis*) population? A case study from south-western Germany. Salamandra. 2013;49: 97–104.

Hajibabaei M, Shokralla S, Zhou X, Singer GAC, Baird DJ. Environmental barcoding: a next-generation sequencing approach for biomonitoring applications using river benthos. PLoS ONE. 2011;6: e17497. pmid:21533287

Thomsen PF, Kielgast J, Iversen LL, Wiuf C, Rasmussen M, Gilbert MTP, et al. Monitoring endangered freshwater biodiversity using environmental DNA. Mol. Ecol. 2012;21: 2565–2573. pmid:22151771

Franklinos LH, Lorch JM, Bohuski E, Fernandez JRR, Wright ON, Fitzpatrick L, et al. Emerging fungal pathogen Ophidiomyces ophiodiicola in wild European snakes. Sci. Rep. 2017;7(1): 3844. pmid:28630406

Stuart SN, Chanson JS, Cox NA, Young BE, Rodrigues ASL, Fischman DL, et al. Status and trends of amphibian declines and extinctions worldwide. Science (New York, N.Y.). 2004;306(5702): 1783–1786.

Vitt LJ, Caldwell JP. Herpetology: An Introductory Biology of Amphibians and Reptiles. San Diego: Academic Press; 2014. pp. 757.

Araújo MB, Thuiller W, Pearson RG. Climate warming and the decline of amphibians and reptiles in Europe. J. Biogeogr. 2005;33: 1712–1728.

Köhler J, Vieites DR, Bonett RM, García FH, Glaw F, Steinke D, et al. New amphibians and global conservation: a boost in species discoveries in a highly endangered vertebrate group. BioScience. 2005;55(8): 693–696.

Ceballos G, Ehrlich PR, Dirzo R. Biological annihilation via the ongoing sixth mass extinction signaled by vertebrate population losses and declines. Proc. Nati. Acad. Sci. 2017;114(30):

Cox NA, Temple HJ. European Red List of Reptiles. Luxembourg: Office for Official Publications of the European Communities. 2009.

Martel A, Blooi M, Adriaensen C, Van Rooij P, Beukema W, Fisher MC, et al. Recent introduction of a chytrid fungus endangers Western Palearctic salamanders. Science. 2014;346: 630–631. pmid:25359973

Wake SD, Vredenurg VT. Are we in the midst of the sixth mass extinction? A view from the amphibians. Proc. Nat. Acad. Sci. U.S.A. 2008;105: 11466–11473.

Sinervo B, Mendez-De-La-Cruz F, Miles DB, Heulin B, Bastiaans E, Villagrán-Santa Cruz M, et al. Erosion of lizard diversity by climate change and altered thermal niches. Science. 2010;328(5980): 894–899. pmid:20466932

Böhm M, Collen B, Baillie JE, Bowles P, Chanson J, Cox N, et al. The conservation status of the world's reptiles. Biol. Conserv. 2013;157: 372–385.

The Council of the European Communities (1992) Council Directive 92/43/EEC of 21 May 1992 on the conservation of natural habitats and of wild fauna and flora (http://eur-lex.europa.eu/legal-content/EN/TXT/?uri=CELEX:31992L0043).

Hebert PDN, Cywinska A, Ball SL, deWaard JR. Biological identifications through DNA barcodes. Proc. Royal Soc. B. 2004;270: 313–321.

Perl RB, Nagy ZT, Sonet G, Glaw F, Wollenberg KC, Vences M. DNA barcoding Madagascar's amphibian fauna. Amphibia-Reptilia. 2014;35(2): 197–206.

Vences M, Nagy ZT, Sonet G, Verheyen E. DNA barcoding amphibians and reptiles. DNA Barcodes. Humana Press, Totowa, NJ. 2012: 79–107.

Smith MA, Poyarkov NA Jr, Hebert PD. DNA barcoding: *CO1* DNA barcoding amphibians: take the chance, meet the challenge. Mol. Ecol. Res. 2008;8(2): 235–246.

Vasconcelos R, Montero-Mendieta S, Simó-Riudalbas M, Sindaco R, Santos X, Fasola M, et al. Unexpectedly high levels of cryptic diversity uncovered by a complete DNA barcoding of reptiles of the Socotra Archipelago. PLoS One. 2016;11(3): pmid:26930572

Chambers EA, Hebert PD. Assessing DNA barcodes for species identification in North American reptiles and amphibians in natural history collections. Plos One. 2016;11(4): pmid:27116180

Ratnasingham S, Hebert PDN. BOLD: The Barcode of Life Data System (www. barcodinglife.org). Mol. Ecol. Notes. 2007;7: 355–364. pmid:18784790

Dejean T, Valentini A, Miquel C, Taberlet P, Bellemain E, Miaud C. Improved detection of an alien invasive species through environmental DNA barcoding: the example of the American bullfrog *Lithobates catesbeianus*. J. Appl. Ecol. 2012;49(4): 953–959.

Hawlitschek O, Nagy ZT, Berger J, Glaw F. Reliable DNA barcoding performance proved for species and island populations of Comoran squamate reptiles. PLoS One. 2013;8(9): e73368. pmid:24069192

Cabela A, Grillitsch H, Tiedemann F. Atlas zur Verbreitung und Ökologie der Amphibien und Reptilien in Österreich: Auswertung der Herpetofaunistischen Datenbank der Herpetologischen Sammlung des Naturhistorischen Museums in Wien. 2001: 1–880.

Gollmann G. Rote Liste der in Österreich gefährdeten Lurche (Amphibia) und Kriechtiere (Reptilia). In: BUNDESMINISTERIUM FÜR LAND- UND FORSTWIRTSCHAFT (Hrsg.): Rote Listen gefährdeter Tiere Österreichs, Teil 2: Kriechtiere, Lurche, Fische, Nachtfalter, Weichtiere. Böhlau Verlag, Wien-Köln-Weimar; 2007. P. 515.

Grabher M, Niederer W. Der Fadenmolch *Lissotriton helveticus* (Razoumowsky, 1789), eine neue Amphibienart für Österreich. UMG Berichte 7. 2011. Retrieved from http://www.umg.at/umgberichte/UMGberichte7_Fadenmolch_2011.pdf

Kindler C, Chèvre M, Ursenbacher S, Böhme W, Hille A, Jablonski D, et al. Hybridization patterns in two contact zones of grass snakes reveal a new Central European snake species. Sci. Rep. 2017;7(1): 7378. pmid:28785033

Plötner J. Möglichkeiten und Grenzen morphologischer Methoden zur Artbestimmung bei europäischen Wasserfröschen (*Pelophylax esculentus*-Komplex). Zeitschrift für Feldherpetologie. 2010;17: 129–146

Speybroeck J, Beukema W, Bok B, Van Der Voort J. Field Guide to the Amphibians & Reptiles of Britain and Europe. Bloomsbury Natural History. 2016. pp. 432.

Richlen ML, Barber PH. A technique for the rapid extraction of microalgal DNA from single live and preserved cells. Mol. Ecol. Notes. 2005;5: 688–691.

Koblmüller S, Salzburger W, Obermüller B, Eigner E, Sturmbauer C, Sefc KM. Separated by sand, fused by dropping water: habitat barriers and fluctuating water levels steer the evolution of rock-dwelling cichlid populations. Mol. Ecol. 2011;20: 2272–2290. pmid:21518059

Duftner N, Koblmüller S & Sturmbauer C. Evolutionary relationships of the Limnochromini, a tribe of benthic deepwater cichlid fish endemic to Lake Tanganyika, East Africa. J. Mol. Evol. 2005;60: 277–289. pmid:15871039

Tamura K, Stecher G, Peterson D, Filipski A, Kumar S. MEGA6: Molecular Evolutionary Genetics Analysis version 6.0. Mol. Biol. Evol. 2013;30: 2275–2279.

Kimura M. A simple method for estimating evolutionary rates of base substitutions through comparative studies of nucleotide sequences. J. Mol. Evol. 1980;16: 111–120. pmid:7463489

Bergsten J, Englund M, Erricson P. A DNA key to all Swedish vertebrates. Forthcoming.

CHAPTER XV:

PHYLUM—CHORDATA—SUBPHYLUM, VERTEBRATA—CLASS, AVES "OUR FEATHERED GRANDPARENTS SEND THEIR LOVE"

"Today our duty is to destroy the myth of evolution, considered as a simple, understood, and explained phenomenon which keeps rapidly unfolding before us. ... The deceit is sometimes unconscious, but not always, since some people, owing to their sectarianism, purposely overlook reality and refuse to acknowledge the inadequacies and falsity of their beliefs." Pierre-Paul Grasse

The animals belonging to the class Aves or birds as is generally known were thought to have evolved about 200 million years ago during the Jurassic period. Instead of being ectothermic, they have become endothermic which means that they are able to maintain a constant body temperature, regardless of their environment. The body is covered by feathers, and horny scales are only found on the feet. The skeletal bones have become light and pneumatic and these modified features are adaptations to flight. The eyes are well developed and this is reflected in the development of an enhanced visual center in the brain. Their inner ear contains a cochlea which is coiled. The narrow jaws have become developed into a horn-covered beak. There are over 8,800 species of birds in existence. Because of their adaptation to flight, they have become more dispersed by inhabiting every terrestrial habitat on earth. They are found in the Polar Regions, at the equator, in deserts, the forest, the jungles, the oceans and the mountains. For this study the pigeons belonging to the order Columbia were chosen (see Figure 15.1).

Figure 15.1: Picture of Parrots.

Body Structure and Movement

The birds' bodies are covered with feathers, which are outgrowths of the integument and are keratinized dead cells. Pigments are usually deposited on the feathers and these pigments are responsible for the array of beautiful colors exhibited by birds. The feathers are usually laid in tracts and do not grow uniformly throughout the whole body. The down feathers cover most of the body, and are used mostly for insulation. The bristles are found around the eyes; and, the contour feathers or flight feathers cover most of the body which are found more on the posterior border of the arm, hand and tail. In between the contour feathers are the filoplumes with nerve endings and these functions as sensory organs for flight movement. The skeleton has become light, thin, hollow, and the lungs have extended into the bones. This process is referred to as pneumatic which are features for adaptations by birds for flight. The skull has developed a large cranial region with large orbits and toothless beaks. The neck region is very long with modification of the cervical region to allow more movement of the head, which is used in conjunction with the beak, for feeding, preening of the feathers, nest building and defense. The trunk region is shortened, and the trunk vertebrae have been modified to become firmly united to act as a fulcrum for the action of the wings, and a strong point of attachment for the pelvic girdle and hind legs. The development of ribs for the protection of the thoracic regions is flexible and allows for respiratory movements. The sternum, or breast plate is another developed features that is broad and do have a large mid-ventral keel, which increases the surface area for the attachment of flight muscles. The fore limbs have become developed into wings to carry the flight feathers for flying. The hind legs have been modified for running, walking and jumping before flight. Since flights by birds require a lot of energy, they have become endothermic and the rate of feeding has been greatly increased to meet this high-energy requirement.

Nervous System

The sensory organs in the birds have become modified to help them with their adaptations to flight. The olfactory organs are less developed. The eyes are large and their sense of sight is well developed. Birds could distinguish objects with different colors, different sizes, and have the ability to accurately judge the correct distance of an object. All these features help in flight, feeding and defense of the animal. The sense of hearing is well developed and helps birds to hear songs and other sounds made by other birds for mating, feeding and other social calls. The birds have large brains with enhanced cerebrum, optic lobes and cerebellum. The brain, in conjunction with the spinal cord, help the bird in behavioral activities in terms of learning, feeding, walking running, jumping, flying, vocalization, courtship, mating, nesting and other social activities.

Circulation, Respiration and Excretion

The circulatory system of birds is well developed. The heart has four completely separated chambers, which allow the complete separation of oxygenated and de-oxygenated blood. The pressure of the blood from the heart is high and has developed large blood vessels to supply the flight muscles. All these enhanced features are modifications that help the birds' adaptations to flight. Further, the birds' adaptations to flight have resulted in an efficient respiratory system with the lungs developing air sacks which acts as bellows and could be extended to the bones and other parts of the body. There is a one-way flow of air in the lungs in contrast to the bi-directional flow in other terrestrial vertebrates. There are anterior and posterior air sacks which play crucial roles during periods of inspiration and expiration. There are two cycles of inspiration and expiration which allow the efficient use of the parabronchi [a group of small parallel branching passages branching from the bronchus] in the absorption of more oxygen into the blood capillaries. The extension of air sacks into the bones and other parts of the body could be controlled by the lungs which allow the body's weight to be decreased or increased for flight.

Birds have metanephric kidneys and these become the principle organs of excretion. The urinary bladder is missing in the birds. The kidneys have been modified to develop large number of kidney tubules with some of them having loops of Henle for more water re-absorption. The development of large number of kidney tubules is as a result of the high metabolic activities of birds, which require more wastes to be excreted. The nitrogenous wastes are excreted in the form of crystals of uric and acid through the cloaca after more water re- absorption at the cloaca (see Figure 15.2).

Figure 15.2: Picture of Birds' mating call.

Digestion

Due to their high requirement of energy for flight and other metabolic activities, birds feed constantly on high-energy foods like fruits, seeds, organic foods, insects, arthropods and small vertebrates. Food taken through the mouth passes through the esophagus to the crop, which acts as food storage and softener. It then moves to the proventriculus where peptic enzymes help break down the food particles. These then pass to the gizzard where the food is mixed with gastric juices and grinded. In most species, stone pellets, bones and other substance are regurgitated from here and not allowed to pass into the small intestine. The almost completely digested food is passed to the small intestine where the bile and pancreatic enzymes complete the digestive processes. Digested food wastes are absorbed here and undigested food or feces are sent out of the body through the cloaca.

Reproduction

Birds are oviparous and have developed cleidoic eggs. Fertilization is internal and some species still have copulatory organs. The birds have lost the right ovary and oviduct, which is an adaptation to reduction of weight for flight. Birds have evolved a complex mating behavior. The males do prepare a nest and exhibit brilliant plumage and colorful songs to attract mates of the same species and ward off other males. They also exhibit territorial spacing to prevent overcrowding. Once a territory had been established and the female has taken residence, and intimate courtship develops. Copulation takes place by the adherence of the male and female cloacae. Sperms are transferred and fertilization occurs internally. The female lays the fertilized eggs in the nest and brood and incubate these eggs until they hatch. In some species, both male and female brood the eggs. Some species do have only the female which brood the eggs while the male bring food to the females. Once the young birds hatch, the females continue to brood and feed them until they are strong enough to walk, run and fly. They also, protect them from potential predators and help them develop other social activities.

Experiments with Mutagens

The adult form, the newly hatched form, and the germ cells of the pigeon were exposed to three mutagens with various intensities ranging from low intensity to high intensity. Ultraviolet rays, x-rays, and radioactive elements were the chosen mutagens. The study utilized the mutagens in the laboratory with the aim of observing their effects on the pigeons, and to ascertain if the mutagens were capable of creating enough mutations to cause spontaneous, or gradual evolution. The cultures were exposed to three various intensities of the mutagens for periods of one month, three month, six months, one year, and three years.

Under low intensity of the mutagens, it was observed that the pigeons actually thrived. Metabolic activities were slightly higher. Nutrition was rapid. Excretion was continuous. The reproductive activities and courtship appeared normal. There were no mutant varieties from developing germ cells.

Under medium intensity of the mutagens, it was observed that the bird initially became very hyperactive. This reaction was more pronounced in the cultures exposed singularly, or in combination to x-rays and radioactive elements. The bird gradually decreased metabolic activities. Flight activities became gradually reduced and the animal did not thrive. Reproductive activities became non- existent. The birds and newly hatched forms did not survive past a couple of months. The germ cells with exposure over a year to the mutagens were artificially fertilized and about 10% of them hatched. But, every one of them had deformities ranging from, loss of wings, loss of hind limbs, loss of eyes, loss of body feathers, reduction in body size and loss of some internal organs. None of the deformed pigeons' chicks survived past two weeks.

Under high intensity of the mutagens, the pigeons did not survive past the fifth day. We could not get the fertilized germ cells to hatch from any of the exposed cultures.

Genetic Mutations

Cells from each culture were excised to observe any genetic mutations occurring within the nuclei. It was observed that there was no change in the chromosomal content. The genome was constant. The total number of genes remained unaffected. However, from the deformed chicks it was observed that there were instances of reduced chromosomal content, but the total numbers of genes per chromosome were unaffected. Further, it was observed that a new organelle associated with the chromosomes, which was termed the MasterCodon, was preventing genotypic mutations.

Conclusion

From this experiment, it was observed that mutations occurring seem to be point or frame shift mutation. The genetic codes that code different proteins seem to be responsible for the phonotypical changes observed in the deformed chicks. Since the total number of genes remained unaffected, the mutagens did not affect the genetic profile of the animal. Rather than change this genetic profile, the animal always succumbs to death. The organelle responsible for the resistance to genetic profile changes is the MasterCodon. The effect of the MasterCondon on cells and ultimately the organism is to allow mutations that will cause favorable adaptations to the environment, but stop all mutations that will alter the genetic profile, or uniqueness of the organism by shutting down the whole cell system resulting in death of the cell or organism.

Suggested Reading

Barnosky AD, Matzke N, Tomiya S, Wogan GO, Swartz B, Quental TB, et al. Has the Earth's sixth mass extinction already arrived? Nature. 2011; 471(7336): 51–57. pmid:21368823

Dirzo R, Young HS, Galetti M, Ceballos G, Isaac NJ, Collen B. Defaunation in the Anthropocene. Science. 2014; 345(6195): 401–406. pmid:25061202

Maxwell SL, Fuller RA, Brooks TM, Watson JEM. Biodiversity: The ravages of guns, nets and bulldozers. Nature. 2016; 536(7615): 143–145. pmid:27510207

WWF. Living Planet Report-2018: Aiming Higher. Gland: Grooten M and Almond REA (Eds). 2018; https://wwf.panda.org/knowledge_hub/all_publications/living_planet_report_2018/. Accessed July, 2019.

Foley JA, DeFries R, Asner GP, Barford C, Bonan G, Carpenter SR, et al. Global consequences of land use. Science. 2005; 309(5734): 570–574. pmid:16040698

Ceballos G, Ehrlich PR, Barnosky AD, García A, Pringle RM, Palmer TM. Accelerated modern human–induced species losses: Entering the sixth mass extinction. Sci Adv. 2015; 1(5): e1400253. pmid:26601195

Fjeldså J, Álvarez MD, Lazcano JM, Leon B. Illicit crops and armed conflict as constraints on biodiversity conservation in the Andes region. AMBIO: A Journal of the Human Environment. 2005; 34(3): 205–211.

Sekercioglu CH. Increasing awareness of avian ecological function. Trends Ecol Evol. 2006; 21(8): 464–471. pmid:16762448

Whelan CJ, Şekercioğlu ÇH, Wenny DG. Why birds matter: from economic ornithology to ecosystem services. J Ornithol. 2015; 156(1): 227–238.

Mace GM, Collar NJ, Gaston KJ, Hilton-Taylor C, Akçakaya HR, Leader-Williams N, et al. Quantification of extinction risk: IUCN's system for classifying threatened species. Conserv Biol. 2008; 22: 1424–1442. pmid:18847444

Rodrigues A, Pilgrim JD, Lamoureux JF, Hoffman M, Brooks TM. The value of the IUCN Red List for conservation. Trends Ecol Evol. 2006; 21(2): 71–76. pmid:16701477

Farrier D, Whelan R, Mooney C. Threatened species listing as a trigger for conservation action. Environ Sci Policy. 2007; (10): 219–229.

Butchart SHM, Stattersfield AJ, Bennun LA, Shutes SM, Akcakaya HR, Baillie JEM, et al. Measuring global trends in the status of biodiversity: Red List Indices for birds. PLoS Biol. 2004; 2: 2294–2304.

Butchart SHM, Stattersfield AJ, Bennun LA, Shutes SM, Akcakaya HR, Baillie JEM, et al. Using Red List Indices to measure progress towards the 2010 target and beyond. Phil. Trans R Soc Lond B. 2005; 1454: 255–268.

Butchart SHM, Akçakaya HR, Chanson J, Baillie JEM, Collen B, Quader S, et al. Improvements to the Red List Index. PLoS One. 2007; 2: e140. pmid:17206275

Xu H, Tang X, Liu J, Ding H, Wu J, Zhang M, et al. China's progress toward the significant reduction of the rate of biodiversity loss. BioScience. 2009; 59(10): 843–852.

Butchart SHM, Walpole M, Collen B, Van Strien A, Scharlemann JP, Almond RE, et al. Global biodiversity: indicators of recent declines. Science. 2010; 328(5982): 1164–1168. pmid:20430971

Szabo JK, Butchart SH, Possingham HP, Garnett ST. Adapting global biodiversity indicators to the national scale: A Red List Index for Australian birds. Biol Conserv. 2012; 148: 61–68.

Butchart SHM. Red List Indices to measure the sustainability of species use and impacts of invasive alien species. Bird Conserv. Int. 2008; 18(S): S245–S262.

Regan EC, Santini L, Ingwall-King L, Hoffmann M, Rondinini C, Symes A, et al. Global trends in the status of bird and mammal pollinators. Conserv Lett. 2015; 8: 397–403.

McGowan PJK, Mair L, Symes A, Westripp J, Wheatley H, Butchart SHM. Tracking trends in the extinction risk of wild relatives of domesticated species to assess progress against global biodiversity targets. Conserv Lett. 2018: e12588.

Young RP, Hudson MA, Terry AMR, Jones CG, Lewis RE, Tatayah V, et al. Accounting for conservation: using the IUCN Red List Index to evaluate the impact of a conservation organization. Biol Conserv. 2014; 180: 84–96.

United Nations. The Sustainable Development Goals Report 2018. New York: United Nations Publications. 2018. https://unstats.un.org/sdgs/files/report/2018/The SustainableDevelopmentGoalsReport2018-EN.pdf. Accessed 7 December 2019.

Lewis OT, Senior MJ. Assessing conservation status and trends for the world's butterflies: the Sampled Red List Index approach. J. Insect Conserv. 2011; 15: 121–128.

Brummitt NA, Bachman SP, Griffiths-Lee J, Lutz M, Moat JF, Farjon A, et al. Green plants in the red: A baseline global assessment for the IUCN sampled Red List Index for plants. PLoS One. 2015; 10(8): e0135152. pmid:26252495

Baillie JE, Collen B, Amin R, Akcakaya HR, Butchart SH, Brummitt N, et al. Toward monitoring global biodiversity. Conserv. Lett. 2008; 1(1): 18–26.

Bubb PJ, Butchart SHM, Collen B, Dublin HT, Kapos V, Pollock C, et al. IUCN Red List Index–Guidance for National and Regional Use. 2009. https://www.iucn. org/content/iucn-red-list-index-guidance-national-and-regional-use-version-11. Accessed February, 2019.

International Union for Conservation of Nature (IUCN) and BirdLife International (BLI). Metadata for Sustainable Development Goal Indicator 15.5.1: Red List Index. 2017. https://unstats.un.org/sdgs/metadata/files/Metadata-15-05-01.pdf. Accessed February, 2019

Hoffmann M, Brooks TM, Butchart SHM, Gregory RD, McRae L. Trends in biodiversity: vertebrates. In: DellaSala DA, Goldstein MI, editors. Encyclopedia of the Anthropocine, vol 3. Oxford: Elsevier. 2017. pp. 175–184.

Saiz JCM, Lozano FD, Gómez MM, Baudet ÁB. Application of the Red List Index for conservation assessment of Spanish vascular plants. Conserv Biol. 2015; 29(3): 910–919. pmid:25580521

Rodrigues ASL, Brooks TM, Butchart SHM, Chanson J, Cox N, Hoffmann M, et al. Spatially explicit trends in the global conservation status of vertebrates. PLoS One. 2014; 9: e113934. pmid:25426636

BirdLife International. State of the world's birds 2018: Taking the pulse of the planet. Cambridge, UK: BirdLife International. 2018. https://www.birdlife.org/sites/default/files/attachments/BL_ReportENG_V11_spreads.pdf. Accessed March, 2019.

BirdLife International [Internet]. Country profile: Colombia. 2018. [accessed 23 November 2018]. http://www.birdlife.org/datazone/country/colombia.

Asociación Colombiana de Ornitología checklist committee 2018 [Internet]. Species lists of birds for South American countries and territories: Colombia. Version 31/July/2018]. [accessed 31 July 2018]. http://www.museum.lsu.edu/~Remsen/ SACCCountryLists.htm

Avendaño JE, Bohórquez CI, Rosselli L, Arzuza-Buelvas D, Estela FA, Cuervo AM, et al. Lista de chequeo de las aves de Colombia: Una síntesis del estado del conocimiento desde Hilty & Brown (1986). Ornitología Colombiana. 2017; 16: 1–83.

McMullan M. Field Guide to the Birds of Colombia. 1st ed. Bogotá: Rey Naranjo Editores; 2018.

Myers N, Mittermeier RA, Mittermeier CG, Da Fonseca GA, Kent J. Biodiversity hotspots for conservation priorities. Nature. 2000; 403(6772): 853–858. pmid:10706275

Stattersfield AJ, Crosby MJ, Long AJ, Wege DC. Endemic Bird Areas of the World. Priorities for biodiversity conservation. Cambridge: BirdLife Conservation Series 7; 1998.

Hazzi NA, Moreno JS, Ortiz-Movliav C, Palacio RD. Biogeographic regions and events of isolation and diversification of the endemic biota of the tropical Andes. Proc Natl Acad Sci. USA 2018; 115(31): 7985–7990. pmid:30018064

N. IUCN Red List Categories and Criteria: Version 3.1. Second edition. Gland, Switzerland and Cambridge: IUCN. 2012. https://portals.iucn.org/library/sites/ library/files/documents/RL-2001-001-2nd.pdf Accessed February, 2019.

IUCN. Guidelines for Application of IUCN Red List Criteria at Regional Levels: Version 3.0. IUCN Species Survival Commission. Gland, Switzerland and Cambridge: IUCN. 2003. https://portals.iucn.org/library/sites/library/files/ documents/RL-2003-001-EN.pdf. Accessed February, 2019.

Hilty SL, Brown WL. A guide to the birds of Colombia. 1 st edition. Princeton: Princeton University Press; 1986.

Ayerbe F. Guía Ilustrada de la Avifauna Colombiana. Wildlife Conservation Society. Primera Edición. Bogotá: Puntoaparte Bookvertising; 2018.

Schulenberg TS. [Internet]. Neotropical Birds Online. 2018. [accessed March 2019]. https://neotropical.birds.cornell.edu/Species-Account/nb/home.

Grubb PJ. Interpretation of the 'Massenerhebung' effect on Tropical Mountains. Nature 1971; 229: 44–45. pmid:16059069

Kattan GH, Alvarez-López H, Giraldo M. Forest fragmentation and bird extinctions: San Antonio eighty years later." Cons Biol. 1994; 8(1): 138–146.

Renjifo LM. Composition changes in a subandean avifauna after long-term forest fragmentation. Cons Biol. 1999; 13(5): 1124–1139.

Stratford JA, Stouffer PC. Forest fragmentation alters microhabitat availability for Neotropical terrestrial insectivorous birds. Biol Conserv. 2015; 188: 109–115.

BirdLife International. [Internet]. IUCN Red List for birds. 2018 [accessed 14 November 2018]. http://www.birdlife.org.

Aide TM, Grau HR. Globalization, migration, and Latin American ecosystems. Science. 2004; 305: 1915–1916. pmid:15448256

CHAPTER XVI:

PHYLUM–CHORDATE–SUBPHYLUM, VERTEBRATE–CLASS, MAMMALIA. "OUR SISTERS ARE HERE TO STAY!"

"I think we need to go further than this and admit that the only acceptable explanation is creation. I know this is an anathema to physicists, as indeed it is to me, but we must not reject a theory that we do not like if the experimental evidence supports it." H. S. Lipson

The animals belonging to the class Mammalia were thought to have evolved about 70 million years ago in the cretaceous period. These animals are endothermic and do have hair and subcutaneous fat as insulating layers. The dermis does have cutaneous glands which secrete sweat, oil and pheromones. The animals have about 4,500 existing species. Except for whales, most of the animals are terrestrial. For this study, the common house rat belonging to the order Rodentia was utilized (see Figure 16.1).

Figure 16.1: Picture of Rodents.

Body Structure and Movement

The fact that rodents are endothermic has helped the animals to develop an active life style that led to modification of the body structure. The skull encloses a large, synaptic braincase and has a well-developed jaw joint that is placed between the dentary bone of the lower jaw and the temporal (squamosal) bone of the skull. The middle ear has three (3) auditory ossicles and there is a spiral cochlea in the inner ear. The cerebrum is large with a gray cortex and divided into large cerebellar hemispheres. The teeth are heterodont and a precise occlusion has been developed. Generally, the teeth replacement is limited to about two or three times in the animal's life cycle. The limbs are usually carried in position beneath the body. Movement is by walking, hopping, or running. For adaptation to an active terrestrial life style, the rodents have developed an arched back which is a fusion of the posterior inclination of the spines at the lumbar vertebrae. The elbow and the knee been shifted close to the trunk in such a way that the elbow points posteriorly and the knee anterior. All these modifications enable the animal to have better mechanical support, and increased stride with greater speed for locomotion.

Circulation, Respiration and Excretion

The rodent heart is completely divided into four chambers which prevents the mixing of oxygenated blood and non-oxygenated blood. This complete separation has allowed for the modification of the ventricles for pumping blood under high pressure to other parts of the body. Due to the active life style of rodents which requires greater gas exchange, they have developed pulmonary alveoli which have increased the surface area of the lungs for more active absorption of oxygen. The development of the diaphragm has also increase the efficiency of lung ventilation. Also, because of the high metabolic activity of animal, a larger amount of nitrogenous wastes are formed and are constantly being expelled from the body. The nitrogenous wastes are formed as urea, and due to its toxicity, a large among of water is needed to expel it from the body. And as such, the animal has developed a large number of kidney tubules glomeruli, and a high blood pressure which allow for high filtration rate and constant removal of water and urea from the blood. The highly developed and efficient loops of Henle allow the re-absorption of over 99% of the water back into the circulatory system. The wastes, in form of urine, is highly concentrated and usually expelled out of the body through the urethra

Nutrition

Rats have to maintain a high level of metabolic activity and as a result of this; they have to consume large quantities of food. They have developed large jaws and musculature associated with nutrition. The food is taken through the mouth where it is grinded, crushed and cut by the teeth in the sockets of the jaws. The heterodont teeth have become modified for varied functions. Digestion of the food starts in the mouth with the secretions of amylase in the salivary glands which helps in the breakdown of starch. The lubricated food is passed to the stomach which helps in the breakdown of starch. The lubricated food is passed to the intestines where peptic enzymes, pancreatic enzymes and bile helps to break down the food further into a form that is easily absorbed through the microscopic linings of the walls in the intestine. Undigested food wastes are passed through the large intestine to the rectum where the feces are expelled through the anus.

Reproduction

Rodents have developed mammary glands, which arises from either sweat or sebaceous glands. The mammary glands (breasts) are unique and are used by the animals for feeding their young, hence, the name Mammalia. Except for a few species, most mammals are viviparous and retain their embryos in the uterus. The number of young ones produced has been reduced due to the fact that the uterus is limited to the amount of embryos it can adequately hold and nurture. The male and female species are separate. The mammals have developed intricate courtship behavior and sexual activity involving intercourse that leads to internal fertilization. Most mammals have developed placentae which are used for the protection and nourishment of the fetus in the uterus. The young mammals are birthed after being fully developed. The mother still feed and nourishes the young mammals through the mammary glands until they are strong enough to fend for themselves (see Figure 16.2).

Figure 16.2: Picture of a Squirel feeding.

Experiment with Mutagens

The adult rat, the embryonic stages and the germ cells were exposed to various intensities of ultraviolet rays, radioactive elements and x-rays. The experiment was primary meant to observe the effect of mutagens in creating mutations in the animal sufficient to create a spontaneous evolution or gradual evolution. For the mutations to be quantifiable, the animals were exposed to the mutagens for periods of one month, three months, six months, one year, and three years.

Under low intensity of the mutagens, the animal continued thriving. There were no observed phenotypical, or genetically changes. The metabolic activity appears normal and there were no significant physiological changes in the animal.

Under medium intensity, increased metabolic activity was observed. The animal became more active and restless. Then, it started withdrawing and remained dormant for longer periods at a time. The animal did not survive past five months. The germ cells developed into embryonic stages with lot of abnormalities. Albinism was rampant. There were some with loss of eyes, limbs, claws, ears, other organs; and a few had deformities with the digestive system. None of the abnormal fetus survived past the first eight days.

Under high intensity, the animal, including the fetal stages and germ cells, did not survive past the first five days. There were no observable genetic mutations, or phenotypical changes since the animal did not survive for these observations to be quantifiable.

Genetic Mutation

From the observations of the cultures, mutations taking place were point and frame shift mutations. The resultant effect of the mutations was to create different codes that produced different proteins and enzymes resulting in different messages and production of abnormal body parts. Nonetheless, the genome, the gene size and the chromosomal number of the animal remained unchanged and constant. It was observed that a new organelle associated with the DNA strand, which was named the MasterCodon, was responsible for maintaining the genotype and protecting the uniqueness of the animal so that it will not evolve from its class to another class or subphylum or phylum.

Conclusion

From the observations, the mutations taking place were not sufficient to create a spontaneous evolution. Observations did show that gradual evolution was also resisted by the animal. Any attempt to mutate to create a new organism was always aborted and the resultant organism never survived. Mutations to adapt the organism to its environment are generally allowed as it does not affect radical changes in the organisms' genotype. The organelle controlling the mutations from becoming chaotic is the MasterCodon. The MasterCodon exert influence on the animal by allowing mutations that are favorable to the organism and will allow the animal to adapt favorably to its environment. However, the MasterCodon maintains the constancy and uniqueness of the organism from generation to generation ensuring smooth transfer of similar genetic material from parent to offspring; and, at the same time maintaining the uniqueness of each of the offspring and simultaneously maintaining the constancy of the family or the species.

Any attempt by the animal to evolve into different class or phylum results in the MasterCodon shutting down the whole animal's organic system or resulting in death of the animal. The MasterCodon seems programed to control life span and the eventual death of the organism.

Suggested Reading

C. Albert, G.M. Luque, F. Courchamp. The twenty most charismatic species.
PloS One, 13 (2018), Article e0199149, 10.1371/journal.pone.0199149

E.S. Bakker, J.F. Pagès, R. Arthur, T. Alcoverro. Assessing the role of large
herbivores in the structuring and functioning of freshwater and marine angiosperm
ecosystems.
Ecography, 39 (2016), pp. 162-179, 10.1111/ecog.01651

J.R. Bennett, R. Maloney, H.P. Possingham. Biodiversity gains from efficient use of
private sponsorship for flagship species conservation.
Proc. R. Soc. Lond. B Biol. Sci., 282 (2015), p. 20142693, 10.1098/rspb.2014.2693

BirdLife International and Handbook of the Birds of the World. Bird species
distribution maps of the world. Version 2018.1.
[WWW Document] http://datazone.birdlife.org/species/requestdis (2018) accessed
2.13.19

B.W. Bowen, J. Roman. Gaia's handmaidens: the Orlog model for conservation
biology.
Conserv. Biol., 19 (2005), pp. 1037-1043, 10.1111/j.1523-1739.2005.00100.x

J.F. Brodie. Is research effort allocated efficiently for conservation? Felidae as a
global case study.
Biodivers. Conserv., 18 (2009), pp. 2927-2939, 10.1007/s10531-009-9617-3

T.M. Brooks, R.A. Mittermeier, G.A.B. da Fonseca, J. Gerlach, M. Hoffmann, J.F.
Lamoreux, C.G. Mittermeier, J.D. Pilgrim, A.S.L. Rodrigues. Global biodiversity
conservation priorities. Science, 313 (2006), pp. 58- 61, 10.1126/science.1127609
(80-.)

F.T. Brum, C.H. Graham, G.C. Costa, S.B. Hedges, C. Penone, V.C. Radeloff, C.
Rondinini, R. Loyola, A.D. Davidson. Global priorities for conservation across
multiple dimensions of mammalian diversity. Proc. Natl. Acad. Sci. U. S. A., 114
(2017), pp. 7641-7646, 10.1073/pnas.1706461114

L. Büchi, S. Vuilleumier. Coexistence of specialist and generalist species is shaped
by dispersal and environmental factors. Am. Nat., 183 (2014), pp. 612-624,
10.1086/675756

G. Burin, W.D. Kissling, P.R. Guimarães, C.H. Şekercioglu, T.B. Quental.
Omnivory in birds is a macroevolutionary sink.
Nat. Commun., 7 (2016), p. 11250, 10.1038/ncomms11250

S. Chamberlain. Rredlist: "IUCN" Red List Client, R Package Version 0.1.0.
(2016)

S. Chamberlain, E. Szoecs, C. Boettiger. taxize: taxonomic search and phylogeny
retrieval.
F1000Research, 2 (2012), p. 191, 10.12688/f1000research.2-191.v1

A.S.A. Chapman, V. Tunnicliffe, A.E. Bates. Both rare and common species make
unique contributions to functional diversity in an ecosystem unaffected by human
activities. Divers. Distrib., 24 (2018), pp. 568-578, 10.1111/ddi.12712

V. Chillo, R.A. Ojeda. Mammal functional diversity loss under human-induced
disturbances in arid lands. J. Arid Environ. (2012), 10.1016/j.jaridenv.2012.06.016

J. Clavel, R. Julliard, V. Devictor. Worldwide decline of specialist species: toward a
global functional homogenization?
Front. Ecol. Environ., 9 (2011), pp. 222-228, 10.1890/080216

A. Colléony, S. Clayton, D. Couvet, M. Saint Jalme, A.C. Prévot. Human
preferences for species conservation: animal charisma trumps endangered status.
Biol. Conserv., 206 (2017), pp. 263-269, 10.1016/j.biocon.2016.11.035

R.S.C. Cooke, A.E. Bates, F. Eigenbrod. Global trade-offs of functional redundancy
and functional dispersion for birds and mammals.
Global Ecol. Biogeogr., 28 (2019), pp. 484-495, 10.1111/geb.12869

R.S.C. Cooke, F. Eigenbrod, A.E. Bates. Projected losses of global mammal and
bird ecological strategies. Nat. Commun., 10 (2019), p. 2279, 10.1038/s41467-019-
10284-z

R.S.C. Cooke, T.C. Gilbert, P. Riordan, D. Mallon. Improving generation length
estimates for the IUCN Red List. PloS One, 13 (2018), Article e0191770, 10.1371/
journal.pone.0191770

M. Davis, S. Faurby, J.-C. Svenning. Mammal diversity will take millions of years
to recover from the current biodiversity crisis.
Proc. Natl. Acad. Sci. U. S. A. (2018), 10.1073/pnas.1804906115
201804906

F. de Mendiburu. Agricolae: statistical procedures for agricultural research.
R Packag. version 1.2.8
https://doi.org/10.1525/california/9780520268326.003.0002 (2017)
S. Dray, A.-B. Dufour. The ade4 package: implementing the duality diagram for
ecologists. J. Stat. Software, 22 (2007), pp. 1-20, 10.18637/jss.v022.i04

J.E. Duffy. Biodiversity and ecosystem function: the consumer connection.
Oikos, 99 (2002), pp. 201-219, 10.1034/j.1600-0706.2002.990201.x

J. Dunning. CRC Handbook of Avian Body Masses.
(second ed.), CRC Press. CRC Press (2008), 10.1017/S0963180113000479

J.A. Estes, J. Terborgh, J.S. Brashares, M.E. Power, J. Berger, W.J. Bond, S.R.
Carpenter, T.E. Essington, R.D. Holt, J.B.C. Jackson, R.J. Marquis, L. Oksanen,
T. Oksanen, R.T. Paine, E.K. Pikitch, W.J. Ripple, S.A. Sandin, M. Scheffer, T.W.
Schoener, J.B. Shurin, A.R.E. Sinclair, M.E. Soulé, R. Virtanen, D.A. Wardle.
Trophic downgrading of planet earth. Science, 333 (2011), pp. 301-306, 10.1126/
science.1205106 (80-.)

D.P. Faith. Conservation evaluation and phylogenetic diversity.
Biol. Conserv., 61 (1992), pp. 1-10, 10.1016/0006-3207(92)91201-3

S.A. Fritz, O.R.P. Bininda-Emonds, A. Purvis. Geographical variation in predictors
of mammalian extinction risk: big is bad, but only in the tropics.
Ecol. Lett., 12 (2009), pp. 538-549, 10.1111/j.1461-0248.2009.01307.x

L. Gamfeldt, H. Hillebrand, P.R. Jonsson. Multiple functions increase the
importance of biodiversity for overall ecosystem functioning.
Ecology, 89 (2008), pp. 1223-1231, 10.1890/06-2091.1

K.J. Gaston Common ecology
Bioscience, 61 (2011), pp. 354-362, 10.1525/bio.2011.61.5.4

K.J. Gaston. Rarity. Population and Community Biology Series (first ed.), Springer,
Netherlands (1994), 10.1007/978-94-011-0701-3

CHAPTER XVII:

COULD MAN HAVE EVOLVED FROM APES?

"We have no acceptable theory of evolution at the present time. There is none; and I cannot accept the theory that I teach to my students each year. Let me explain. I teach the synthetic theory known as the neo-Darwinian one, for one reason only; not because it's good, we know it is bad, but because there isn't any other. Whilst waiting to find something better you are taught something which is known to be inexact, which is a first approximation. . ."
Professor Jerome Lejeune

The apes, belonging to the order primate, were supposed to have evolved about 60 million years ago in the Eocene period. Man belonging to the same order primates was supposed to have evolved about three million years ago in the Pleistocene period. Evolutionists believe that both man and apes evolved from a common ancestry from a primitive eutherian stock. They refer to similarities in there metabolic characteristic as evidence that man evolved form apes However, fossil records have not been available, or discovered, that will link apes to man (see Figure 17.1).

Figure 17.1: A depiction of Ape gradual evolution to Human.

HERE IS WHAT WE KNOW

Man and apes do have flat faces with a stereoscopic vision and the ability to sit on their haunches and examine object with their hands. They have higher mental functions that allows for symbolization and conceptual thought which is an adaptation from an enlarged brain. Both of the groups have lost their tails and could assume an upright posture.

Apes, like man, do exhibit complex behaviors by living in social groups with an organized hierarchy which is led by a dominant male. They could communicate by facial expressions and primitive sounds with the ability to use tools like sticks for gathering food (see Figure 17.2).

Figure 17.2: A picture of Gorillas.

However, despite these similarities, man differs from apes in a lot of ways. Man is better adapted to walking on two legs (bipedal) and has the ability to stand upright with better gait than apes. Human has lost the grasping function which is still present in apes. Generally, apes are stronger and taller than man. Man uses his longer thumb and the shorter metacarpal portion of his hands for gripping, holding, carrying and making tools, or objects. The hands are not used for locomotion as in apes. The body hair in man is very sparse and the dermis does contain an increased number of sweat glands relative to apes. Also, the human brain is larger and more complex that that of ape. Further, man's parietal, frontal, and temporal areas of the cerebral cortex are much enlarged that that of apes which enables man to have better cognitive thinking, speech abilities, conceptual thought, and organized and active memory.

The reproduction and development patterns in man are much different relative to that of apes. Female apes only copulate with males during the few days of estrus around the middle of their ovarian cycle. Human females, however, can copulate at most times including pregnancy and lactation periods. Apes' birth is simple in that the fetus's head is smaller than the pelvic canal of the female. At birth, the young ape is relatively more mature and within a few days of birth could move easily about and fend for itself. Unlike in the human females, birth could be difficult in that the head of the fetus is bigger than the female pelvic canal. At birth, the human fetus is immature and will be cared for by the parents over a long period of time.

WHAT WE DON'T KNOW

Despite the apparent similarities between man and apes, scientists still do not know precisely how and when man evolved from apes. Fossil records from Egypt dating back to 30million years ago did reveal that early primitive apes did exist then. It is believed that the pongids (Great Apes) did evolve from the early apes and were found in Africa as early as 15million years ago.

The earliest Hominids (Human) were represented by Australopithecus Africanus as far back as 3.7 million years ago and were believed to have evolved from the pongids. Australopithecus Robustus and Australopithecus Afarensis were believed to have evolved from Australopithecus Africanus about 2.5 million years ago. None of these species survived past 1.5 million years ago. The modern Hominids represented by Homo erectus did evolve from the Australopithecus about 2.0 million years ago. Modern man or, Homo sapiens were first believed to have evolved from Homo erectus about 300,000 years ago. The greatest problem facing the evolutionists is the absence of any fossil records of any ape-like or, Hominid-like animal that revealed the gradual progression or evolution of the hominid from apes in the past 5 million years.

Is it possible that such link does not exist? Is it possible that the hominid just appeared suddenly?

Despite these mysteries, man and apes do share close genetical structure. The human genome size is about 3.3×10^9 bp while the ape's genome size is about 3.2×10^9 bp. With such close genetic structure, why don't apes develop into humans? And, why don't humans develop into apes?

Could the MasterCodon associated with the DNA be responsible for this constancy, or permanency? Could this organelle be responsible for maintaining the uniqueness of man and that of apes?

In the final chapter of this book, it will be shown conclusively, that man could not have evolved from ape, and current theories of evolution are substantially faulty.

Suggested Reading

Alba DM (2012). Fossil apes from the Vallès-Penedès Basin. Evolutionary Anthropology 21: 254–269.

Alba DM, Fortuny J, Moyà-Solà S (2010a). Enamel thickness in the middle Miocene great apes Anoiapithecus, Pierolapithecus and Dryopithecus. Proceedings of the Royal Society B 277: 2237–2245.

Alba DM, Fortuny J, Perez de los Rios M, Zanolli C, Almécija S, Casanovas-Vilar I, Robies JM, Moyà-Solà S (2010b). New dental remains of Anoiapithecus and the first appearance datum of hominoids in the Iberian Peninsula. Journal of Human Evolution 65: 573–584.

Almécija S, Alba DM, Moyà-Solà S (2009). Pierolapithecus and the functional morphology of Miocene ape hand phalanges: paleobiological and evolutionary implications. Journal of Human Evolution 57: 284–297.

Almécija S, Alba DM, Moyà-Solà S, Köhler M (2007). Orang-like manual adaptations in the fossil hominoid Hispanopithecus laietanus: first steps towards great ape suspensory behaviours. Proceedings of the Royal Society B 274: 2375–2384.

Almécija S, Smaers JB, Jungers WL (2015). The evolution of the human and ape hand proportions. Nature Communications 6: 7717.

Almécija S, Tallman M, Alba DM, Pina M, Moyà-Solà S, Jungers WL (2013). The femur of Orrorin tugenensis exhibits morphometric affinities with both Miocene apes and later hominins. Nature Communications 4: 2888.

Alpagut B, Andrews P, Fortelius M, Kappelman J, Temizsoy I, Celebi H, Lindsay W (1996). A new specimen of Ankarapithecus meteai from the Sinap Formation of central Anatolia. Nature 382: 349–351.

Alpagut B, Andrews P, Martin L (1990). New hominoid specimens from the middle Miocene site at Paşalar, Turkey. Journal of Human Evolution 19: 397–422.

Andrews P (1978). A revision of the Miocene Hominoidea of East Africa. Bulletin of the British Museum of Natural History (Geology) 30: 85–224.

Andrews P (1990). Palaeoecology of the Miocene fauna from Paşalar, Turkey. Journal of Human Evolution 19: 569–582.

Andrews P (1992). Evolution and environment in the Hominoidea. Nature 360: 641–646.

Andrews P (1995). Ecological apes and ancestors. Nature 376: 555–556.

Andrews P (1996). Palaeoecology and hominoid palaeoenvironments. Biological Reviews 71: 257–300.

Andrews P (2015). An Ape's View of Human Evolution. Cambridge, Cambridge University Press.

Andrews P, Bamford M (2008). Past and present vegetation ecology of Laetoli, Tanzania. Journal of Human Evolution 58: 78–98.

Andrews P, Cameron D (2010). Rudabanya: taphonomic analysis of a fossil hominid site from Hungary. Palaeogeography, Palaeoclimatology, Palaeoecology 297: 311–329.

Andrews P, Cronin J (1982). The relationships of Sivapithecus and Ramapithecus and the evolution of the orang utan. Nature 297: 541–546.

Andrews P, Harrison T (2005). The last common ancestor of apes and humans. In Interpreting the Past: Essays on Human, Primate, and Mammal Evolution in Honor of David Pilbeam (Lieberman DE, Smith RJ, Kelley J, eds.), pp 103–121. Boston, Brill Academic Publishers.

Andrews P, Tobien H (1977). New Miocene locality in Turkey with evidence on the origin of Ramapithecus and Sivapithecus. Nature 268: 699–701.

Andrews P, Van Couvering JH (1975). Palaeoenvironments in the East African Miocene. In Approaches to Primate Paleobiology (Szalay FS, ed.), pp 62–103. Basel, Karger.

Andrews P, Walker AC (1976). The primates and other fauna from Fort Ternan, Kenya. In Human Origins (Isaac G, ed.), pp 279–304. Menlo Park, Benjamin.

Andrews P, Lord J, Evans EMN (1979). Patterns of ecological diversity in fossil and modern mammalian faunas. Biological Journal of the Linnean Society 11: 177–205.

Arcadi AC, Wrangham RW (1999). Infanticide in chimpanzees: review of cases and a new within-group observation from the Kanyawara study group in Kibale National Park. Primates 40: 337–351.

Axelrod DI (1975). Evolution and biogeography of the Madrean-Tethyan sclerophyll vegetation. Annals of the Missouri Botanical Garden 62: 280–334.

Axelrod DI (2000). A Miocene (10–12 Ma) Evergreen Laurel-Oak Forest from Carmel Valley, California. University of California Publications in Geological Sciences 145. Berkeley, University of California Press.

Axelrod DI, Raven PH (1978). Cretaceous and Tertiary vegetation history in Africa. In Biogeography and Ecology of Southern Africa (Werger MJA, ed.), pp 77–130. The Hague, Junk.

Barrett L, Dunbar RIM, Lycett J (2002). Human Evolutionary Psychology. Basingstoke, Palgrave.

Beard KC, Teaford MF, Walker A (1986). New wrist bones of Proconsul africanus and P. nyanzae from Rusinga Island, Kenya. Folia Primatologia 47: 97–118.

Begun DR (1992). Miocene fossil hominids and the chimp-human clade. Science 257: 1929–1933.

Begun DR (1993). New catarrhine phalanges from Rudabánya (Northeastern Hungary) and the problem of parallelism and convergence in hominoid postcranial morphology. Journal of Human Evolution 24: 373–402.

Begun DR (1994). Relations among the great apes and humans: new interpretations based on the fossil great ape Dryopithecus. Yearbook of Physical Anthropology 37: 11–63.

Begun DR (2002). European hominoids. In The Primate Fossil Record (Hartwig WC, ed.), pp 339–368. Cambridge, Cambridge University Press.

Begun DR (2009). Dryopithecins, Darwin, de Bonis and the European origin of the African apes and human clade. Geodiversitas 31: 789–816.

Begun DR (2013). The Miocene hominoid radiations. In A Companion to Paleoanthropology (Begun DR, ed.), pp 398–416. Oxford, Wiley-Blackwell.

Begun DR, Güleç E (1998). Restoration of the type and palate of Ankarapithecus meteai: taxonomic and phylogenetic implications. American Journal of Physical Anthropology 105: 279–314.

Begun DR, Kivell TL (2011). Knuckle-walking in Sivapithecus? The combined effects of homology and homoplasy with possible implications for pongine dispersals. Journal of Human Evolution 60: 158–170.

Begun DR, Kordos L (1993). Revision of Dryopithecus brancoi Schlosser, 1910, based on the fossil hominid material from Rudabánya. Journal of Human Evolution 25: 271–286.

Begun DR, Geraads D, Guleç E (2003). The Çandır hominoid locality: implications for the timing and pattern of hominoid dispersal events. Courier Forschungsinstitut Senckenberg 240: 251–265.

Begun DR, Nargolwalla MC, Kordo L (2012). European Miocene hominids and the origin of the African ape and human clade. Evolutionary Anthropology 21: 10–23.

Begun DR, Teaford MF, Walker A (1994). Comparative and functional anatomy of Proconsul phalanges from the Kaswanga primate site, Rusinga Island, Kenya. Journal of Human Evolution 26: 89–165.

Bestland E (1990). Sedimentology and paleopedology of Miocene alluvial deposits at the Paşalar hominoid site, western Turkey. Journal of Human Evolution 19: 363–377.

Bestland EA, Retallack GI (1993). Volcanically influenced calcareous palaeosols from the Miocene Kiahera Formation, Rusinga Island, Kenya. Journal of the Geological Society 150: 293–310.

Beynon AD, Dean MC, Leakey MG, Reid DJ, Walker A (1998). Comparative dental development and microstructure of Proconsul teeth from Rusinga Island, Kenya. Journal of Human Evolution 35: 163–209.

Blumenschine R (1987). Characteristics of an early hominid scavenging niche. Current Anthropology 28: 383–407.

CHAPTER XVIII:

THE THEORY OF ABSOLUTISM

"It is not the strongest of species that survives, nor the most intelligent that survives. It is the one that is the most adatable to change." Charles Darwin

The origin of life which has been shrouded in mysteries has been challenging to most scientists and religious leaders leading to various speculations in attempts to explain this phenomenon.

Evolution theories are mostly speculations which have been legitimatized by scientists using cell theories and Darwinian Theories to justify their hypotheses with regards to the origin life.

According to Darwin, the motor of evolution operates by modifying what is already available. **In order words, evolution does not build organism from scratch**. Therefore, how was the original cell created and from what? It is believed that heritable variation is responsible for evolution which occurs as a result of constant mutations and recombination and these fuel evolutions.

Biology has its foundation built on these evolution theories.

Query: Are these theories factual?

There is no doubt that cell and molecule theories are factual; but, their application to evolutionary trends is questionable.

From our experiments and other observations, this author was able to prove in this final chapter that evolution did not take place, and Darwin's Theories of evolution is moot.

In 1945, during the Second World War II, atomic bombs were dropped on Nagasaki and Hiroshima. There were a lot of radioactive elements in the atmosphere and these affected the people of these cities in varying ways. Majority died from the radiations. Majority became deformed and eventually died. Majority had deformed offspring. In all the observed cases, mutation did take place, but they were lethal to the organism and the people affected. There was no observed change in the human species or other animals, or, organisms in those cities. In other words, mutation though strong, could not, or, did not affect evolutionary trends in such a way as to create new organisms

Conclusive Experiments

It is correct that Darwin and other scientists advocated that evolutionary process is slow and take place over long period of time.

Our experiments were carried out to determine the adverse effect of natural occurring mutagens (sun rays, ultra – violet rays, x – rays and naturally occurring radioactive chemicals) on different organisms in the animal kingdom which were carried out for period of five years. The experiments were carried out to determine if mutations occurred in the organisms due to the effect of the mutagens.

We did conclude that mutations did occur, but only to help the organisms adapt to their environment. Mutations that are extreme will only result in the death of the organism. Extreme mutations, regardless of how strong the intensity of mutagens applied to the organisms, did not modify the organisms into new order or class or phylum.

It was observed that organisms are virtually constant within species in terms of genome, gene size and chromosomal number. Essentially, adverse mutation did not change the cellular or, molecular structure within the organisms in each species that will create a new organism. Point mutations, or, shift mutation that occurred, were only phenotypical and meant to allow the organism to adapt to its changing environment. In other words, an organism will have its whole system shut down, or killed off, if there was an attempt by extreme mutations to change the organism from its original species to a new species or order or class or phylum.

The cellular organelle responsible for maintaining the constancy in organism **is the MasterCodon. This genetic organelle in the nucleus of each organism is responsible for the uniqueness of each organism within each species and will shut down, or kill the organism, if there is an attempt for an organism to change from one species to a new species or new order or new class or phylum.**

If the MasterCodon was not exerting this control, then, our universe will have gone crazy and there will be chaos – Evolution would have gone crazy!!! Essentially, for every day we wake up there will be millions of new animal and plant species or orders or classes or phyla!! In fact, you may wake up and see that your spouse in bed with you has evolved into a new and different species from you!! Your dogs and cats would have changed into new organisms, or, animals belonging to different orders!! In other words, evolution would have gone crazy! We will then consistently have **''EVOLVING CREATURES''** roaming our universe!

The MasterCodon prevents this chaos from taking place in our universe. For there to be a genuine evolution, species will have to change into genus, then, into an order, then a sub-class, then a class and next a phylum

If Darwin's theories and the evolutionists were correct, then, we should constantly and everyday wake up to new species of animals and plants.

For example, the cockroaches which have been on this planet for over 300 million years would have evolved into totally new organisms. They are still here. They have not changed into new organisms. They are still being cockroaches. The MasterCodon in the nucleus of this organism has allowed it to maintain its constancy.

Most fish have been in existence for over 500 million years. They are still fish. There have been no changes to the organism. They did not develop into new organisms. They have remained relatively constant with the exception of modifications to their environment. The MasterCodon still gives the fish their uniqueness and help them to maintain their uniqueness.

From protozoans to mammals, we have observed in history, that, many of these animals came in existence in different geological times. It is possible that fossil records are not quite accurate because most fossils of organisms cannot be recovered. But, we are certain that these organisms did not evolve drastically into new animals. The cataclysm events that occurred in history, did not create new organisms, but, rather, resulted in extinction of existing organisms. These findings support the theory that the MasterCodon will kill, or shut down the organism, rather than allow it to change into a new organism.

From our findings genuine evolution could not have taken place. The MasterCodon within the nucleus of each organism exerts a lot of control on the organism in maintaining each species unique and constant characteristics. The same MasterCodon controls the organism's life span, specific coloring, specific characteristics, specific voices, or sounds, specific eyes, specific shapes, specific and unique fingerprints, etc. (See Table 18.1).

Table 18.1: Summary of effects of Mutagens on different Organisms.

Animals	Experiments	Mutations	Evolution
Protozoa, Paramecium	Organism exposed to mutagens (radioactive chemicals, alpha-rays, ultra-violet rays, and x-rays) for periods of one month to 3 years.	Observed few mutants varieties. The mutants did not survive. Phenotypical changes seen. Under High intensity of the mutagens, the organisms did not survive past 5 days.	The organism did not evolove or change to another organism. No changes in genotype.
Porifera, Red Boring Sponge	Organism exposed to mutagens (radioactive chemicals, alpha-rays, ultra-violet rays, and x-rays) for periods of one month to 3 years.	Observed few mutants varieties. The mutants did not survive. Phenotypical changes seen. Under High intensity of the mutagens, the organisms did not survive past 5 days.	The organism did not evolove or change to another organism. No changes in genotype.
Cnidaria, Hydra	Organism exposed to mutagens (radioactive chemicals, alpha-rays, ultra-violet rays, and x-rays) for periods of one month to 3 years.	Observed few mutants varieties. The mutants did not survive. Phenotypical changes seen. Under High intensity of the mutagens, the organisms did not survive past 5 days.	The organism did not evolove or change to another organism. No changes in genotype.
Platyhelminthes, Tapeworm	Organism exposed to mutagens (radioactive chemicals, alpha-rays, ultra-violet rays, and x-rays) for periods of one month to 3 years.	Observed few mutants varieties. The mutants did not survive. Phenotypical changes seen. Under High intensity of the mutagens, the organisms did not survive past 5 days.	The organism did not evolove or change to another organism. No changes in genotype.
Aschelminthes, Roundworm	Organism exposed to mutagens (radioactive chemicals, alpha-rays, ultra-violet rays, and x-rays) for periods of one month to 3 years.	Observed few mutants varieties. The mutants did not survive. Phenotypical changes seen. Under High intensity of the	The organism did not evolove or change to another organism. No changes in genotype.

		mutagens, the organisms did not survive past 5 days.	
Mollusca, Snail	Organism exposed to mutagens (radioactive chemicals, alpha-rays, ultra-violet rays, and x-rays) for periods of one month to 3 years.	Observed few mutants varieties. The mutants did not survive. Phenotypical changes seen. Under High intensity of the mutagens, the organisms did not survive past 5 days.	The organism did not evolove or change to another organism. No changes in genotype.
Annelida, Earthworm	Organism exposed to mutagens (radioactive chemicals, alpha-rays, ultra-violet rays, and x-rays) for periods of one month to 3 years.	Observed few mutants varieties. The mutants did not survive. Phenotypical changes seen. Under High intensity of the mutagens, the organisms did not survive past 5 days.	The organism did not evolove or change to another organism. No changes in genotype.
Crustacea, Lobster	Organism exposed to mutagens (radioactive chemicals, alpha-rays, ultra-violet rays, and x-rays) for periods of one month to 3 years.	Observed few mutants varieties. The mutants did not survive. Phenotypical changes seen. Under High intensity of the mutagens, the organisms did not survive past 5 days.	The organism did not evolove or change to another organism. No changes in genotype.
Uniramia, Cockroach	Organism exposed to mutagens (radioactive chemicals, alpha-rays, ultra-violet rays, and x-rays) for periods of one month to 3 years.	Observed few mutants varieties. The mutants did not survive. Phenotypical changes seen. Under High intensity of the mutagens, the organisms did not survive past 5 days.	The organism did not evolove or change to another organism. No changes in genotype.
Echinodermata, Sea Star	Organism exposed to mutagens (radioactive chemicals, alpha-rays, ultra-violet rays, and x-rays) for periods of one month to 3 years.	Observed few mutants varieties. The mutants did not survive. Phenotypical changes seen. Under High intensity of the mutagens, the organisms did not survive past 5 days.	The organism did not evolove or change to another organism. No changes in genotype.

Pisces, Bony Fish	Organism exposed to mutagens (radioactive chemicals, alpha-rays, ultra-violet rays, and x-rays) for periods of one month to 3 years.	Observed few mutants varieties. The mutants did not survive. Phenotypical changes seen. Under High intensity of the mutagens, the organisms did not survive past 5 days.	The organism did not evolove or change to another organism. No changes in genotype.
Amphibia, Clawed Frog	Organism exposed to mutagens (radioactive chemicals, alpha-rays, ultra-violet rays, and x-rays) for periods of one month to 3 years.	Observed few mutants varieties. The mutants did not survive. Phenotypical changes seen. Under High intensity of the mutagens, the organisms did not survive past 5 days.	The organism did not evolove or change to another organism. No changes in genotype.
Reptilia, Common Lizard	Organism exposed to mutagens (radioactive chemicals, alpha-rays, ultra-violet rays, and x-rays) for periods of one month to 3 years.	Observed few mutants varieties. The mutants did not survive. Phenotypical changes seen. Under High intensity of the mutagens, the organisms did not survive past 5 days.	The organism did not evolove or change to another organism. No changes in genotype.
Aves, Pigeon	Organism exposed to mutagens (radioactive chemicals, alpha-rays, ultra-violet rays, and x-rays) for periods of one month to 3 years.	Observed few mutants varieties. The mutants did not survive. Phenotypical changes seen. Under High intensity of the mutagens, the organisms did not survive past 5 days.	The organism did not evolove or change to another organism. No changes in genotype.
Mammalia, House Rat	Organism exposed to mutagens (radioactive chemicals, alpha-rays, ultra-violet rays, and x-rays) for periods of one month to 3 years.	Observed few mutants varieties. The mutants did not survive. Phenotypical changes seen. Under High intensity of the mutagens, the organisms did not survive past 5 days.	The organism did not evolove or change to another organism. No changes in genotype.

CLONING

Genuine cloning that will change an organism to move to different genus or order or class is virtually impossible as of now. We have observed protozoan cells dividing mitotically. Such divisions of cells in metazoans are not common, except in cancerous cells. Cloning that result from cell division is generally phenotypical. Example, like Dolly the lamb–this is generally a phenotypic cloning whereby the offspring look like or are similar to the parent. In humans, we do observe twins that are identical in facial or exterior appearance given rise to phenotypical resemblance.

It should be noted, that in identical twins and any cloning of cells, the new organisms, or offspring, are uniquely differently from each other in their personality, fingerprints, eye trait and other specific traits unique only to that organism. In other words, we can see the physical resemblance of a cloned lamb, but, its uniqueness is identifiable as in human. The agent creating these differences in hypothetical cloned organism is the MasterCodon. This nucleic structure will allow phenotypical resemblance, but, will ensure that the new offspring have their own individuality and unique characteristics, or traits peculiar to each living organism. However, if the MasterCodon can be manipulated, then true cloning may be possible. In addition, we may create new monsters if the MasterCodon is tampered with.

CELLULAR LONGEVITY

From our experiments, it was observed that in each nucleus of a cell within the organism, the MasterCodon exists within the genes. This organelle controls: (1) the constancy of organism; (2) the organism by resisting any change in the organism from its current state into a new species or genus order; (3) the organism by shutting down cellular system resulting in death; (4) the life span, or longevity of each cell and eventually the organism as a whole entity; (5) the organism giving it, its unique individuality, and specific identifiable characteristics or traits.

With such controlling influence, the MasterCodon could be effectively manipulated to increase life span of each cell, and, eventually the whole organism. It is, therefore, possible that man could live again up to 1,000 years old if the MasterCodon is controlled effectively. It is, also, possible to cure all human diseases if the MasterCodon is effectively manipulated to marshal, or strengthen the human immunological defenses against all diseases, like AIDS, cancer, and heart diseases.

END OF DARWIN AND EVOLUTION THEORIES

It is correct that Darwin's natural selection theory, as it applies to organisms, only affect the organism phenotypes; and, that point and shift mutations occurring will only allow the organism to adapt to it environment. All other evolutionary theories as presented in Chapter 1 of this book are moot based on the findings explicated below in this text.

However, it is incorrect to apply cellular mutations and survival of the fittest to evolution and to deduct that all species evolve from a singular cell. And, that, the **singular** cell arose from a "Big Bang" theory. These assertions **and all other evolution theories are invalid.**

Here are the observable facts that fully disprove Darwin and all other evolution theories:

1) **Whatever happened to evolution? Did evolution stopped?** If we accept evolution theories of spontaneous or gradual evolution, then, **EVOLUTION MUST HAVE STOPPED.** Because if evolution did not stop, then, we should observe new species of animals and plants every day. In other words, there will be billions of new species in our universe daily. There will be chaos and confusion on our planet. Instead of evolving genotypically, organisms will die. **Organisms' order and constancy is maintained by the MasterCodon in every nucleus of every cell in every organism.**

2) If evolution theories were valid, then animals like fish, amphibians, reptiles and protozoans that have been in existence for over 200 million years would continue to evolve and we will observe, daily, new creatures inhabiting our planet.

3) If evolution theories were valid, then, animals like protozoans, fish, amphibians and reptiles by virtue of natural selection and survival of the fittest and the fact that they have been in existence for over 200 million years would have been more highly evolved than man, more intelligent than man, have better speech than man, have better armies, computer and organized structures than man. But, this is not the case. Rather, fish is still fish. Amphibians are still amphibians. Reptiles are still reptiles. And, man is still man, being the domineering and ruling animal of the universe.

4) The genome size of most animals reflects the complexity and state of evolution of the animals. For example, protozoans' genome is about 2.3×10^5. Fish genome is about 3.4×10^8. Amphibians' genome is about 3.1×10^9. And, man genome is about 3.3×10^9. But, it has been observed that some amphibian do have genome size of about 3.1×10^{10}. And some flowering plants have genome size which are more complex, or advance than man. That is not what evolutionists' theories state. Based on genome size, the amphibians and plants should be more intelligent than man. But, this is not the case. Rather, man is still the domineering ruler of the animal kingdom. If evolution was correct, then, we should have newly evolved creatures from the amphibians, or, the flowering plants ruling the world.

5) The genome size of the common toad (amphibian), X. Laevis is 3.1×10^9 and that of man is 3.3×10^9. If evolution was correct, then there is nothing stopping the embryos of the toad from becoming a human zygote since their genome is so close together and there is a need for the toad to move to an advance closest stage in evolution, and in this case, to a man-like creature. However, this is not the case rather, the MasterCodon, in each cellular organism prevents species from changing from its original state to a new species or organism or genus or order or class.

6) If evolution is correct, then man with a genome size of 3.3×10^9 should evolve into a higher toad-like creature with genome size 3.1×10^{10} or into a plant-like creature with a genome size of 4.4×10^{11}. But, this is not the case. Rather, the MasterCodon in each cell of the organism maintain order and constancy in the organism and prevent it from changing into new organism or genus or order or class. The facts are, therefore, against evolution being the theory that best explain the origin and existence of life in our universe.

7) The Sadhill Cranes have been in existence for 10 million years and have not evolved, and, are still being Cranes today. The Lice has been in existence for 20 million years and has not evolved, and, are still being Lice today. The Purple Frogs have been in existence for 50 million years and have not evolved, and, are still being Frogs today. The Echidnas have been in existence for 60 million years and have not evolved, and, are still being Echidnas today.

8) The Crocodiless have been in existence for 75 million years and have not evolved, and, are still being Crocodiles today. The Cow Sharks have been in existence for 80 million years and have not evolved, and, are still being Cow Sharks today. The Giant Freshwater Stingrays have been in existence for 80 million years and have not evolved, and, are still being Stingrays today. The Frilled Sharks have been in existence for 95 million years and have not evolved, and, are still being Frill Sharks today.

9) The Bees have been in existence for 100 million years and have not evolved, and, are still being Bees today. The Platpuses have been in existence for 110 million years and have not evolved, and, are still being Platypuses today. The Ants have been in existence for 120 million years and have not evolved, and, are still being Ants today. The Lizards have been in existence for 200 million years and have not evolved, and, are still being Lizards today. The Tortoises have been in existence for 220 million years and have not evolved, and, are still being Tortoises today.

10) The Horseshoe Shrimps have been in existence for 250 million years and have not evolved, and, are still being Horseshoe Shrimps today. The Cockroaches have been in existence for 260 million years and have not evolved, and, are still being Cockroaches today. The Elephant Sharks have been in existence for 400 million years and have not evolved, and, are still being Elephant Sharks today. The Jelly Fishes have been in existence for 500 million years and have not evolved, and, are still being Jelly Fishes today.

11) The Sea Sponges have been in existence for 600 million years and have not evolved, and, are still being Sea Sponges today. The Cyanobacteria have been in existence for 3.5 billion years and have not evolved, and, are still being Cyanobacteria today.

12) So, what happened to Evolution? Did Evolution stop? The animals listed above and other animals plus all plants are still the **SAME CONSTANT ORGANISMS FOR THE PAST FOUR BILLION YEARS** which never changed and never evolved. These are the facts. **EVOLUTION, IT NEVER HAPPENED. PERIOD**. Therefore, Darwin and Evolutionary Theories are, hereby, disproved. (See Table 18.2).

Table 18.2: List of Animals that never evolved, but remained constant for over millions of years.

Animals	Years in Existence	Evolution
Sandhill Crane	10 million	Still the same organism today. Maintained a constant genotype. Never evolved or changed.
Lice	20 million	Still the same organism today. Maintained a constant genotype. Never evolved or changed.
Purple Frog	50 million	Still the same organism today. Maintained a constant genotype. Never evolved or changed.
Echidna	55 million	Still the same organism today. Maintained a constant genotype. Never evolved or changed.
Crocodile	60 million	Still the same organism today. Maintained a constant genotype. Never evolved or changed.
Cow Shark	60 million	Still the same organism today. Maintained a constant genotype. Never evolved or changed.
Giant Freshwater Stingray	65 million	Still the same organism today. Maintained a constant genotype. Never evolved or changed.
Filled Shark	95 million	Still the same organism today. Maintained a constant genotype. Never evolved or changed.
Bee	100 million	Still the same organism today. Maintained a constant genotype. Never evolved or changed.
Duck-billed Platypus	110 million	Still the same organism today. Maintained a constant genotype. Never evolved or changed.
Green Sea Turtle	110 million	Still the same organism today. Maintained a constant genotype. Never evolved or changed.
Martialis Hureka Ant	120 million	Still the same organism today. Maintained a constant genotype. Never evolved or changed.
Goblin Shark	125 million	Still the same organism today. Maintained a constant genotype. Never evolved or changed.
Tuatara Lizard	200 million	Still the same organism today. Maintained a constant genotype. Never evolved or changed.
Tortoise	200 million	Still the same organism today. Maintained a constant genotype. Never evolved or changed.
Sturgeon	200 million	Still the same organism today. Maintained a constant genotype. Never evolved or changed.
Tadpole Shrimp	220 million	Still the same organism today. Maintained a constant genotype. Never evolved or changed.
Horseshoe Shrimp	250 million	Still the same organism today. Maintained a constant genotype. Never evolved or changed.
Cockroach	300 million	Still the same organism today. Maintained a constant genotype. Never evolved or changed.
Lamprey	360 million	Still the same organism today. Maintained a constant genotype. Never evolved or changed.
Coelacanth	400 million	Still the same organism today. Maintained a constant genotype. Never evolved or changed.

Elephant Shark	400 million	Still the same organism today. Maintained a constant genotype. Never evolved or changed.
Horseshoe Crab	450 million	Still the same organism today. Maintained a constant genotype. Never evolved or changed.
JellyFish	500 million	Still the same organism today. Maintained a constant genotype. Never evolved or changed.
Nautilus	500 million	Still the same organism today. Maintained a constant genotype. Never evolved or changed.
Velvet Worm	520 million	Still the same organism today. Maintained a constant genotype. Never evolved or changed.
Brachiopod	530 million	Still the same organism today. Maintained a constant genotype. Never evolved or changed.
Sea Sponge	760 million	Still the same organism today. Maintained a constant genotype. Never evolved or changed.
Ctenophora	760 million	Still the same organism today. Maintained a constant genotype. Never evolved or changed.
Cyanobacteria	4,000 million	Still the same organism today. Maintained a constant genotype. Never evolved or changed.

THE THEORY OF ABSOLUTISM

Based on our experiments, findings, and observations, I, therefore, propose **Amadasun's Theory of Absolutism or perfect creation**, as follows:

1)   Evolution involving genotypical changes for organisms to evolve from a species, to a genus, to an order, to a class, and to a phylum did not happen and could not have taken place. All existing evolutionary theories are invalid and disproved.

2)   Darwin's Theories on natural selection and the common ancestry of all species is faulty and hereby disproved.

3)   Man did not and could not have evolved from ape or monkey.

4)   Living organisms did not evolve from each other. Natural selection help species through mutations, to adapt to its changing environment, and generally, such mutations are phenotypical in nature.

5)   For there to be a genuine evolution, there should be genotypical changes and there should be species of living organisms constantly changing to another species, and then, to another genus, and then, to another family, and then, to another class, and then, to another phylum, and then, to another or new kingdom.

If evolution is valid, then, we should by now, have more new kingdoms with millions of new known and unknown species. This is not the case. Organisms of various species, genus, family, class, and phylum have remained constant and unchanged for the past 600 million years. Therefore, evolution never took place period.

6) The MasterCodon ensures that species do not change from its original state to a totally new species. The MasterCodon does allow mutations that are favorable to species of living organisms in their adaptations to their environment for their survival. The MasterCodon, also control and maintain the unique characteristics, or traits of every living organism peculiar only to that organism, and thereby ensuring the individuality of each organism in terms of personality, mannerism, fingerprints, eyes differentiations, body shape, coloring, voices, sound, gaits, postures, etc.

7) **If there was no stabilizing gene, or organelle like the MasterCodon, then evolution would have really gone crazy.** In fact, every day we wake up, we will see new creatures on our planet every second, every minute, and every hour. We will observe that our spouse, or children, or parents, or nephews, or nieces, or friends, or pets (dogs and cats,) have changed into completely new creatures – **we will continuously have EVOLVING CREATURES every day.**

We will then observe that every day there are continuously "Evolving Creatures" walking and living around us. There will have been chaos in our universe. However, this is not the case because the MasterCodon prevent such crazy evolution taking place.

8) Whenever there is fertilization between organisms of the same species, every offspring produced is different from each other and from their parents and from other organisms based on the MasterCodon replicating itself differently in a mathematical function as follows:

$$T = \infty$$
And,
$$T = (A^{46} \times B^{46})\ ^{(M = 200,000)\ (K = 20)\ (G = 3,000,000,000)}$$

Where,

T = Total number of combinations possible in any zygote formation determining the uniqueness and individuality of each organisms.

A = Pair of chromosomes from the material side of the organism = 46.

B = Pair of chromosomes from the paternal side of the organism = 46.

K = the amount amino acids combinations possible = 20.

G = the amount of nucleotides combinations possible = 3,000,000,000.

M = 100,000 x 2 = the total amount of genetic materials present in the MasterCodon of each parent.

It will be observed that in any given combination, or zygote formation, T approaches infinity, and as such the possibility of any organism being completely alike to another organism in terms of individuality and uniqueness is infinity.

Suggested Reading

Blumenschine RJ, Cavallo JA (1988). Scavenging and human evolution. Scientific American 267: 90–96.

Boesch C (1996). Three approaches for assessing chimpanzee culture. In Reaching into Thought (Russon A, Bard K, Parker S, eds.), pp 404–429. Cambridge, Cambridge University Press.

Boesch C, Boesch H (1990). Tool use and tool making in wild chimpanzees. Folia Primatologica 54: 86–99.

Boesch C, Boesch-Achermann H (2000). The Chimpanzees of the Tai Forest: Behavioural Ecology and Evolution. Oxford, Oxford University Press.

Brown B, Ward SC (1988). Basicranial and facial topography in Pongo and Sivapithecus. In Orang Utan Biology (Schwartz J, ed.), pp 247–260. New York, Plenum Press.

Brunet M, Guy F, Pilbeam D, Zollikofer C (2002). A new hominid from the Upper Miocene of Chad, Central Africa. Nature 418: 145–151.

Casanovas-Vilar I, Alba DM, Moyà-Solà S, Galindo J, Cabrera L, Garces M, Furia M, Robles JM, Köhler M, Angelone C (2008). Biochronological, taphonomical and paleoenvironmental background of the fossil great ape Pierolapithecus catalaunicus (Primates, Hominidae). Journal of Human Evolution 55: 589–603.

Cerling TE (1992). Development of grasslands and savannas in East Africa during the Neogene. Palaeogeography, Palaeoclimatology Palaeoecology 97: 241–247.

Cerling TE, Harris JM, Ambrose S, Leakey MG, Solounias NJ (1997). Dietary and environmental reconstruction with stable isotope analyses of herbivore tooth enamel from the Miocene locality of Fort Ternan, Kenya. Journal of Human Evolution 33: 635–650.

Cerling TE, Wynn JG, Andanje SA, Bird MI, Korir DK, Levin NE, Mace W, Macharia AN, Quade J, Remien CH (2011). Woody cover and hominin environments in the past 6 million years. Nature 476: 51–56.

Chaimanee Y, Jolly D, Benanmmi M, Tafforeau P, Duzer D, Moussa I, Jaeger J-J (2003). A middle Miocene hominoid from Thailand and orangutan origins. Nature 422: 61–65.

Chaimanee Y, Suteethorn V, Jintsakul P, Vidthayanon C, Marandat B, Jaeger J-J (2004). A new orang-utan relative from the late Miocene of Thailand. Nature 427: 439–441.

Chesters KI (1957). The Miocene flora of Rusinga Island, Lake Victoria, Kenya. Palaeontographica 101: 30–71.

Collinson ME, Andrews P, Bamford M (2009). Taphonomy of the early Miocene flora, Hiwegi Formation, Rusinga Island, Kenya. Journal of Human Evolution 57: 149–162.

Cook JM, Rasplus JY (2003). Mutualists with attitude: coevolving fig wasps and figs. Trends in Ecological Evolution 18: 241–248.

Darwin C (1859). On the Origin of Species by Means of Natural Selection. London, John Murray.

Dean MC (2006). Tooth microstructure tracks the pace of human life-history evolution. Proceedings of the Royal Society B 273: 2799–2808.

Deane AS (2009). Early Miocene catarrhine dietary behaviour: the influence of the Red Queen effect on incisor shape and curvature. Journal of Human Evolution 56: 275–285.

De Bonis L, Koufos GD (2014). First discovery of postcranial bones of Ouranopithecus macedoniensis (Primates, Hominoidea) from the late Miocene of Macedonia (Greece). Journal of Human Evolution 74: 21–36.

De Bonis L, Melentis J (1977). Un nouveau genre de primate hominoide dans le Vallesien (Miocène supérieur) de Macédoine. Comptes Rendus de l'Académie des Sciences, Paris 284: 1393–1396.

De Bonis L, Melentis J (1978). Les primates hominoides du Miocène supérieur de Macédoine. Annales de Paléontologie 64: 185–202.

De Bonis L, Bouvrain G, Geraads D, Koufos G (1992). Diversity and palaeoecology of Greek late Miocene mammalian faunas. Comptes Rendus de l'Académie des Sciences, Paris 291: 99–121.

Elias RB, Gil A, Silva L, Fernández-Palacios JM, Azevedo B, Reis F (2016). Natural zonal vegetation of the Azores Islands: characterization and potential distribution. Phytocoenologia 46: 107–123.

Ersoy A, Kelley J, Andrews P, Alpagut B (2008). Hominoid phalanges from the middle Miocene site of Paşalar, Turkey. Journal of Human Evolution 54: 518–529.

Fernandez-Jalvo Y, Andrews P (2011). When humans chew bones. Journal of Human Evolution 60: 117–123.

Foley RA (2001). The evolutionary consequences of increased carnivory in hominids. In Meat-Eating and Human Evolution (Stanford CB, Bunn HT, eds.), pp 305–331. Oxford, Oxford University Press.

Fuss J, Spassov N, Begun DR, Bohm M (2017). Potential hominin affinities of Graecopithecus from the late Miocene of Europe. PLoS One 12: e0177127.

Galik K, Senut B, Pickford M, Gommery D, Treil J, Kupervage AJ, Eckhardt RB (2004). External and internal morphology of the BAR 1002'00 Orrorin tugenensis femur. Science 305: 1450–1453.

Gebo DL, Beard KC, Teaford MF, Walker A, Larson SG, Jungers WL, Fleagle JG (1988). A hominoid proximal humerus from the early Miocene of Rusinga Island, Kenya. Journal of Human Evolution 17: 393–401.

Gebo DL, Malit NR, Nengo IO (2009). New proconsuloid postcranials from the early Miocene of Kenya. Primates 50: 311–319.

Goodall J (1986). The Chimpanzees of Gombe: Patterns of Behavior. Cambridge, Harvard University Press.

Groves CP (1989). A Theory of Human and Primate Evolution. Oxford, Oxford Scientific Publications.

Groves CP (2001). Primate Taxonomy. Washington, The Smithsonian Institution.

Haile-Selassie Y (2001). Late Miocene hominids from the Middle Awash. Nature 412: 178–181.

Haile-Selassie Y (2004). Late Miocene teeth from Middle Awash, Ethiopia, and early hominid dental evolution. Science 303: 1503–1505.

Harrison RD (2010a). Figs and the diversity of tropical rainforests. Ecography 33: 148–158.

Harrison T (1991). The implications of Oreopithecus for the origins of bipedalism. In Origine(s) de la bipédie chez les hominidés (Coppens Y, Senut B, eds.), pp 235–244. Paris, Cahiers de Paléoanthropologie, CNRS.

Harrison T (1993). Cladistic concepts and the species problem in hominoid evolution. In Species Concepts and Primate Evolution (Kimbel WH, Martin LB, eds.), pp 345–371. New York, Plenum Press.

Harrison T (2002). Late Oligocene to middle Miocene catarrhines from Afro-Arabia. In The Primate Fossil Record (Hartwig WC, ed.), pp 311–338. Cambridge, Cambridge University Press.

Harrison T (2010b). Dendropithecoidea, Proconsuloidea and Hominoidea. In Cenozoic Mammals of Africa (Werdelin L, Sanders WJ, eds.), pp 429–469. Berkeley, University of California Press.

Harrison T (2012). Apes among the tangled branches of human origins. Science 327: 532–534.

Harrison TS, Harrison T (1989). Palynology of the late Miocene Oreopithecus-bearing lignite from Baccinello, Italy. Palaeogeography, Palaeoclimatology, Palaeoecology 76: 45–65.

Harrison T, Rook L (1997). Enigmatic anthropoid or misunderstood ape? The phylogenetic status of Oreopithecus bambolii reconsidered. In Function, Phylogeny, and Fossils: Miocene Hominoid Evolution and Adaptations (Begun DR, Ward CV, Rose MD, eds.), pp 327–362. New York, Plenum Press.

Hartwig WC (2002) (ed.). The Primate Fossil Record. Cambridge, Cambridge University Press.

Heizmann E, Begun DR (2001). The oldest Eurasian hominoid. Journal of Human Evolution 41: 463–481.

Hockings KJ, Anderson AR, Matsuzawa T (2010). Flexible feeding on cultivated underground storage organs by rainforest-dwelling chimpanzees at Bossou, West Africa. Journal of Human Evolution 58: 227–233.

Hughes JF, Skaletsky H, Pynitkova T, Graves TA, van Daalen KM, Minx PJ, Fulton RS, McGrath SD, Locke DP, Friedman C, Trask BJ, Mardis ER, Warren WC, Repping S, Rozen S, Willson RK, Page C (2010). Chimpanzee and human Y chromosomes are remarkably divergent in structure and gene content. Nature 463: 536–539.

Humphrey LT, Andrews P (2008). Metric variation in the postcanine teeth from Pasalar, Turkey. Journal of Human Evolution 54: 503–517.

Ishida H, Pickford M (1997). A new late Miocene hominoid from Kenya: Samburupithecus kiptalami gen. et sp. nov. Comptes Rendus de l'Académie des Sciences, Paris 325: 823–829.

Ishida H, Kunimatsu Y, Nakatsukasa M, Nakano Y (1999). New hominoid genus from the middle Miocene of Nachola. Anthropological Science107: 189–191.

Ishida H, Kunimatsu Y, Takano T, Nakano Y, Nakatsukasa M (2004). Nacholapithecus skeleton from the Middle Miocene of Kenya. Journal of Human Evolution 46: 69–103.

Izquierdo T (2011). Vegetation indices changes in the cloud forest of La Gomera Island (Canary Islands) and their hydrological implications. Hydrological Processes 25: 1531–1541.

Jacobs BF (1987). A middle Miocene (12.2 my old) forest in the East African Rift Valley, Kenya. Journal of Human Evolution 16: 147–155.

Jacobs BF (1992). Taphonomy of a middle Miocene authochthonous forest assemblage, Ngorora Formation, central Kenya. Palaeogeography, Palaeoclimatology, Palaeoecology 99: 31–40.

Kappelman J, Richmond BG, Seiffert ER, Maga AM, Ryan TM (2003). Hominoidea (Primates). In Geology and Paleontology of the Miocene Sinap Formation (Fortelius M, Kappelman J, Sen S, Bernor R, eds.), pp 90–124. New York, Columbia University Press.

Kay RF (1977). Diet of early Miocene African hominoids. Nature 268: 628–630.

Kelley J (2008). Identification of a single birth cohort in Kenyapithecus kizili and the nature of sympatry between K. kizili and Griphopithecus alpani at Paşalar. Journal of Human Evolution 54: 530–537.

Kelley J, Andrews P, Alpagut B (2008). A new hominoid species from the middle Miocene site of Paşalar, Turkey. Journal of Human Evolution 54: 455–479.

www.ingramcontent.com/pod-product-compliance
Lightning Source LLC
Chambersburg PA
CBHW040254100426
42811CB00011B/1265